CHEMICAL BONDING
IN TRANSITION METAL
CARBIDES

CHEMICAL BONDING IN TRANSITION METAL CARBIDES

Alan Cottrell

Department of Materials Science and Metallurgy, University of Cambridge

THE INSTITUTE OF MATERIALS

Book 613
First Published in 1995 by
The Institute of Materials
1 Carlton House Terrace
London SW1Y 5DB

© The Institute of Materials 1995

ISBN 0 901716 68 5

British Library Cataloguing-in-Publication Data

A catalogue record for this book
is available from the British Library

Typeset by Fakenham Photosetting Ltd
Fakenham, Norfolk

Printed and finished in the UK at
The University Press Cambridge

CONTENTS

PREFACE

TRANSITION METAL carbides are scientifically interesting, as well as being industrially important. The affinity of early transition metals such as titanium for carbon leads to compounds with simple crystal structure, high melting point and great hardness. The simplicity of structure is not maintained in the carbides of the later metals.

The superficial clues to the nature of the chemical bonding in these carbides are quite misleading. The NaCl-type structure, which is endemic among the carbides of the early metals, suggests an ionic compound; but they are metallic and the carbon is covalently bonded to the metal. The metallic conductivity, FCC metallic sublattice and variability of composition, in these NaCl-type carbides suggest a metal crystal in which carbon is interstitially dissolved; but in for example TiC there is almost no bonding between the metal atoms, virtually all the valence electrons having gone instead into carbon–metal bonding orbitals.

In recent years, following the development of density functional theory, there have been many calculations of the electronic band structures of these carbides, which have demonstrated the major role played by covalent bonds formed between the p states of the carbon atoms and the valence d states of the metal ones. Inevitably, the computational complexity of such calculations has largely limited their application to the NaCl-type structure and stoichiometric composition. To extend the theory to arbitrarily varied compositions and other structures a simplified model is thus needed.

My aim in this work, which is based on papers I have published recently in *Materials Science and Technology*, has been to seek an essentially qualitative understanding of all such carbides. The method is semi-empirical in that the parameters of its primary equations have been chosen from observed values. In this way it has been possible to set up equations for the cohesive energy which are consistent with the qualitative features of the band theory and yet are flexible enough to be applied, with second-order corrections, to a wide variety of compositions and structures.

ACKNOWLEDGEMENTS

I am grateful to Professor C. Humphreys for making available the facilities of the Department of Materials Science and Metallurgy, University of Cambridge, during the course of this work. It is also a pleasure to acknowledge the encouragement given to me, to write this book, by Mr. G.C. Smith, Dr. J.A. Charles and The Institute of Materials.

Alan Cottrell
December 1994

1

CLOSE-PACKED SPHERE MODELS

1.1 CELLULAR UNITS OF STRUCTURE

FOLLOWING THE method of Frank and Kaspar (1958) it is convenient to view the crystal structures of the transition metal carbides in terms of a cellular unit of structure, i.e. the enclosing cage of neighbouring metal (M) atoms which defines a central ('interstitial') site for a carbon (C) atom. We shall see that there are three such cells and that the various crystal structures differ in their choice of these, in the arrangement of their cells, and in minor deviations from perfectly regular cellular form.

Frank and Kaspar dealt with intermetallic compounds in which the component atoms differed only slightly in size and packed together as closely as possible. With a relatively large atom in the central site it was then inevitable that the number of neighbouring atoms in the surrounding cage would be large, i.e. 12 to 16. On this basis Frank and Kaspar constructed their system of complex intermetallic crystal structures. In the carbides, however, the central C atom is much smaller. Thus, the covalent radius of carbon is 0.077 nm whereas the metallic radius of a transition element ranges from 0.124 (Fe) to 0.160 (Zr) and to even higher values for Y (0.181), La (0.187) and the actinides (e.g. Th, 0.180). The number of equal spheres which could be close-packed round a central one of about only half of their radius is then obviously much smaller, e.g. 6, and this changes the course of the analysis away from that pioneered by Frank and Kaspar.

The same 'close packing' principle of investigating the consequences of maximising this number will be adopted, however. In the intermetallic compounds this principle is obviously justified by the metallic bonding; but the carbon atom is notable for forming more open structures with a small coordination number, $z = 4$. The reason why maximum values of z, compatible with packing equal spheres round a small central one, are to be considered, will be discussed later (chapter 3.3). Several of the crystal structures to be discussed are complex, with metal atom neighbours at slightly different distances. Thus the Frank–Kaspar definition of coordination number will be used; i.e. a Wigner–Seitz polyhedron will be

1

constructed round the atom in question and the number of faces of this will be taken as z. It will be convenient to subdivide this according to the type of neighbouring atoms. Thus, we shall write for an atom of type A, $z_A = z_{AA} + z_{AB}$, where z_{AA} and z_{AB} are its respective numbers of neighbours of types A and B. It will also be convenient to express all interatomic distances as fractions of R, the metal-metal nearest neighbour distance ($R = 2r_m$).

A cellular cage can be made by adding M 'spheres', above and/or below, to a plane unit of M spheres. Only two such plane units are admissible in a regular system, the equilateral triangle, with 3 M spheres, with a central in-plane interstitial site of radius $r_c = 0.155 \, r_m$; and the square (4 M) with $r_c = 0.4142 \, r_m$. On the classical Hägg (1931) basis the central site of a pentagon (5 M) and all larger plane units is already too large, i.e. $r_c = 0.69 \, r_m$ for the pentagon.

The addition of a single M atom, symmetrically above the triangle, creates the smallest three-dimensional cage, the *tetrahedron* with 4 M and $r_c = 0.225 \, r_m$, which is too small. The addition of a second M, on the other side, creates the trigonal bi-pyramid (5 M), i.e. two tetrahedra back to back, with no increase in r_c.

Cells based on the square plane unit are more interesting, because they can give an r_c/r_m near to the carbon/metal ratio. There are four regular such cells (Fig. 1.1):

(i) The *octahedron* or *trigonal antiprism* (6 M) formed by adding 2 M atoms, symmetrically, above and below the square; and consisting of 8 triangular faces. The central site of the square is also that of the octahedron, so that $r_c = 0.4142 \, r_m$.

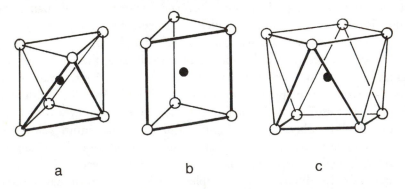

a   b   c

Fig. 1.1 Cellular units of structure with central sites for carbon atoms: (a) octahedron (or trigonal antiprism); (b) trigonal prism; (c) square antiprism.

(ii) The *trigonal prism* (6 M), constructed by forming two more squares on two parallel edges of the original one, so making 3 square and 2 triangular faces. The shifting of the interstitial site from the in-plane centre of the square to the centre of the prism increases the size of this site, $r_c = 0.528 \, r_m$.

(iii) The *square antiprism* (8 M) with 2 parallel squares, rotated 45°relative to one another about a common central axis perpendicular to their plane; and joined by 8 triangles; $r_c = 0.65 \, r_m$.

(iv) The *simple cube* (8 M) with $r_c = 0.732 \, r_m$.

The physical requirement, $r_c \simeq 0.5 \, r_m$, thus leaves only the octahedron, trigonal prism and square antiprism as reasonable candidate cellular units for the carbide crystal structures. In considering the spatial arrangement of many such units, in a crystal, it is useful to deal separately with high-carbon, i.e. $x \simeq 1$ in MC_x, and low-carbon crystals.

1.2 STRUCTURES OF HIGH CARBON CRYSTALS

In the ideal stoichiometric carbide MC, which is closely approached by those of Ti, Zr, Hf, V, Nb, Ta, Mo and W, the number of cells occupied by carbon atoms (or available to be occupied, in the more usual case $MC_{1-\delta}$, where $\delta \ll 1$) is equal to the number of M atoms. It follows that in a crystal in which each C atom has a cage of n neighbours then, for an atomistically uniform structure, every M atom must be a member of n such cages. We are thus interested in $n = 6$ for the octahedron and trigonal prism and $n = 8$ for the square antiprism.

First, consider the possibility of filling space with such cells, as if they were Wigner–Seitz polyhedra. The regular trigonal prism obviously can fill space by forming a simple hexagonal assembly. The efficiency of this space-filling has the effect that there are two cells per M atom, the M atoms themselves forming a simple hexagonal sublattice (with $z_{mm} = 8$). If all these cells were filled, the resulting MC_2 structure would be of the AlB_2 type. But in WC and MoC, which are based on this cell, only one-half of the cells are occupied, in an ordered arrangement such that the carbon atoms also form a simple hexagonal sublattice which interpenetrates the M sublattice. The further effects which confine WC and MoC to only the stoichiometric composition will be discussed later.

The octahedron and square antiprism are not space filling polyhedra and so can be built into crystal structures only with the aid of intermediate, joining, structures. As a result, they cannot provide as many cells per atom as the trigonal prism. The somewhat complex shape of the square antiprism can be accommodated only with the aid of a relatively thick envelope of intermediate material so that, as a result, this structure appears only in carbides with much lower x (i.e. $M_{23}C_6$). But the fitting together of octahedra is much easier, so that many near-stoichiometric carbides based on this cell exist.

The allowed crystal structures built upon the octahedron can be found by adding material to its faces until a space-filling Wigner–Seitz cell is formed. Thus, by adding 8 (non-regular) tetrahedra to its triangular faces the W–S polyhedron of the FCC lattice can be formed. The stacking of these enlarged cells in this lattice then produces the familiar NaCl-type structure, characteristic of the large family of cubic MC carbides, which consists of two interpenetrating FCC sublattices, one for the M atoms, the other for the C (Fig. 1.2). In terms of the edge length, R, of the octahedron the volume of the latter is $0.4714\,R^3$ whereas that of the W–S cell is $R^3/\sqrt{2}$ so that the packing efficiency is 0.667. The octahedra are joined edge-to-edge in pairs, so that each is joined to 12 neighbours, in FCC positions, with intermediate volumes between their separated faces.

By adding suitable other forms to the 8 faces of the octahedron the W–S polyhedron of the BCC lattice could be produced. The volume of this is $(4/3\sqrt{3})\,R^3$, so that the packing efficiency is only 0.612. The structure is not formed.

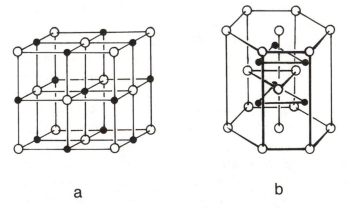

a

b

Fig. 1.2 Crystal structures based on octahedral cellular units for carbon sites: (a) NaCl-type; (b) NiAs-type. Note that in carbides with structure (b) only about one-half of the C-sites are occupied.

The interpretation of the octahedron as a trigonal antiprism, with a trigonal axis directed through the centres of two opposite triangular faces, opens the way to the formation of a hexagonal space-filling structure, made by fitting appropriate non-regular tetrahedra on the 6 side faces of the antiprisms. This leads to a hexagonal crystal structure, of NiAs-type, with a CPH metal sublattice and a simple hexagonal C-site one (Fig. 1.2). As with the NaCl-type structure, it provides one octahedral cell per M atom, with (for an ideal CPH sublattice) the same packing efficiency. However, this structure is not formed in stoichiometric MC carbides, for reasons discussed later.

1.3 STRUCTURES OF LOW CARBON CRYSTALS

In crystals with a high density of octahedral cells, i.e. the FCC and CPH-based ones described above, the move to low carbon contents can be accomplished simply by leaving some of the C-sites empty, in either an ordered or disordered arrangement. This occurs in the cubic carbides, the composition of which extends down to about $MC_{0.5}$ for the group IVA metals and about $MC_{0.6}$ for group VA. At about $MC_{0.5}$ the CPH-based structure appears, as a second phase for the group VA metals and also for molybdenum and tungsten. By contrast the WC and MoC phases, based on the trigonal prismatic cell, do not extend appreciably below the stoichiometric composition, for a reason discussed later.

The move to low carbon content provides an opportunity for the formation of other, more complex, crystal structures which contain a lower density of C-cells and thus require more extensive intermediate structures. The variety of arrangements by which the latter can provide the necessary fitting together then enables the C-cells to take up more complex patterns in relation to one another, patterns determined by various secondary factors.

This development of complex C-cell patterns does not happen with octahedral cells but underlies the structures of many compounds of compositions M_3C_2, M_7C_3, and M_3C, based on the trigonal prismatic cell, as well as $M_{23}C_6$, based on the square prismatic one.

The orthorhombic structure of Cr_3C_2, shown in Fig. 1.3, is built round a 'vertical' axis. One third of the trigonal prismatic cells have their trigonal axis along this vertical direction; the other two-thirds are aligned along a horizontal axis. The structure can be regarded as built from 6-cell units,

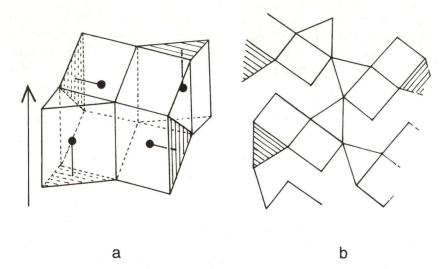

a   b

Fig. 1.3 The M_3C_2 structure: (a) a unit of two vertical and four horizontal trigonal prismatic cells. Metal atoms occupy all cell corners. Carbon atoms are shown at the ends of lines projected into the cells from the striped triangular faces; (b) Pattern of adjoining units, viewed along the vertical axis. The plain and striped triangular faces of vertical cells are at half-cell differences of level along the vertical axis.

stacked in columns along the vertical axis, each such unit consisting of two vertically aligned and four horizontally aligned cells. Every vertical cell contains a carbon atom but only alternate horizontal ones do. The choice of cell occupancy is ordered in a pattern that maximises the distance between neighbouring carbon atoms. The crystal has three crystallographically different M sites, of equal abundance, giving an average metal-metal coordination number, $z_{mm} = 10$, which provides an advantage over that of the WC structure ($z_{mm} = 8$).

The advantage of a high carbon-metal coordination number, $z_{cm} = 8$, is achieved in the square antiprism at the expense of forming a cellular unit with square faces that are awkwardly oriented for building into a crystal structure. The problem is solved, 'ingeniously', by forming a simple accommodation cube of M atoms on one square face and a corresponding cubo-octahedron on the other. The cubic symmetries of these two poly-hedra, with the same orientation of cube axes, enable $M_{23}C_6$ to be thus constructed as a complex cubic crystal (Fig. 1.4). Because the simple cube and the cubo-octahedron are large accommodation structures, only a low

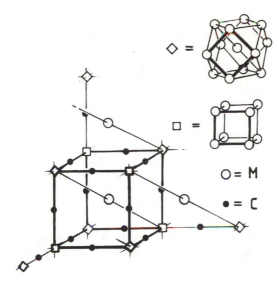

Fig. 1.4 The $M_{23}C_6$ structure.

density of C-cells is possible, so that the carbon content must be low ($x = 0.261$). There are several cystallographically different sites for the M atoms, with a range of R values, giving an average $z_{mm} \simeq 11$. It should be noted that the actual z_{mm} of a given M atom correlates inversely with its z_{mc}. Thus, of the 23 M atoms, 8 have $z_{mc} = 3$ and $z_{mm} = 10$; 12 have $z_{mc} = 2$, $z_{mm} = 11$; 1 has $z_{mc} = 0$, $z_{mm} = 12$; and 2 have $z_{mc} = 0$, $z_{mm} = 16$.

The carbides M_7C_3 and M_3C also provide $z_{cm} > 6$, this time by the access given to external M atoms through the large square faces of their trigonal prismatic C-cells. The complex hexagonal (with an orthorhombic variant, in Cr_7C_3) structure of M_7C_3 is illustrated in Fig. 1.5, based on the analysis of Dyson and Andrews (1969); see also Yakel (1985). It is built in columns, along the hexagonal axis, of empty octahedra, stacked (triangular) face to face, with the hexagonal axis perpendicular to the joined faces. These parallel 'vertical' columns are arranged in a hexagonal pattern and joined by bridging material which includes the important trigonal prismatic C-cells. The latter extend outwards from the triangular side faces of the octahedra, like spokes of a bicycle wheel, and join at their outer triangular faces to vertical columns of trigonal bi-pyramids. Dyson and Andrews showed that the structure can be regarded as being built up from four types of solid figure, seen from the side in Fig. 1.5. Starting in the middle, they are:

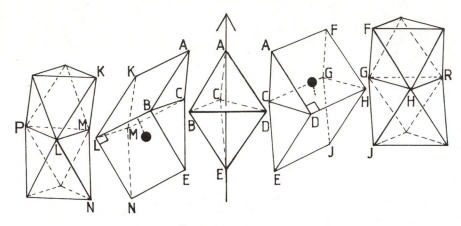

Fig. 1.5 Components of the M_7C_3 structure, in the representation of Dyson and Andrews (1969). Metal atoms occupy all cell corners and the positions of carbon are indicated.

(i) a trigonal bi-pyramid of height $c = $ AE, where c is the height of the crystallographic unit cell along the vertical axis, indicated by the arrow,
(ii) a right trigonal prism, containing a carbon atom, joined on its 'square' face BCML (or CDGH) to an empty skew trigonal prism,
(iii) a similar double prism, upside down with respect to the previous one,
(iv) a column of octahedra, stacked on common triangular faces (e.g. LMP and GHR).

The positions of the C atoms are indicated in these figures; and each vertice represents the site of a metal atom. In Cr_7C_3 each C-site is large enough to accommodate a sphere of radius 0.07 nm, which approximates to the covalent radius of the carbon atom, whereas the empty sites in the octahedra could hold one of radius only 0.055 nm. Another significant feature is that, because of the skewed structure of the empty trigonal prism, a metal atom is sited, e.g. at K, approximately over the centre of the 'square' face, BCML, and fairly close to the carbon atom within. Its distance is about 1.3 of that of a metal atom of the C-cell itself (at B, C, M, L, N and E) to this carbon atom, so that $z_{cm} > 6$.

The various metal atoms again fall into several crystallographically different sets, with slightly different R and z_{mm} values. On average $z_{mm} \simeq 10.7$.

The other structure, based on trigonal prismatic C-cells with $z_{cm} > 6$, M_3C

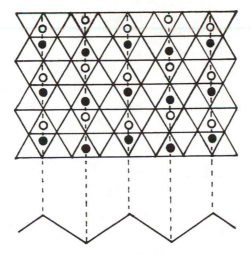

Fig. 1.6 Top and front view of atomic layer of M_3C formed by folding CPH basal planes along their lines of intersection with mirror planes (broken lines). Metal atoms occupy the triangular vertices. Carbon atoms sit on these in positions marked by circles. Empty and full circles are occupied in alternate layers. After Fasiska and Jeffrey (1965).

typified by the cementite of carbon steel, Fig. 1.6, has been usefully analysed by Fasiska and Jeffrey (1965). They pointed out that the description of the structure is simplified by the existence in it of a family of parallel mirror planes. Starting with a hypothetical CPH crystal of iron atoms, the basal planes of this are then sharply bent, alternately up and down to an angle of 112.2°, along their lines of intersection with $(2\,\bar{1}\,\bar{1}\,0)$ planes which then become the mirror planes of the carbide crystal. This folding divides the interstitial positions, along the fold lines, into two sets, as shown in Fig. 1.6, and the carbon atoms occupy these positions in an alternating sequence from one layer to the next. Finally, there are some minor readjustments of neighbouring M positions.

This folding divides the M atoms into two classes; one-third are in positions on the mirror planes, designated 'special' by Fasiska and Jeffrey, and the other two-thirds are in 'general' positions between them. The trigonal prismatic C-cells are not immediately manifest in this representation, but the structure is such that each such cell has two general M atoms, situated centrally just outside two of its 'square' faces, so that $z_{cm} > 6$. As a result, each special M atom has $z_{mm} = 12$ and $z_{mc} = 2$, whereas each general one has $z_{mm} = 11$, $z_{mc} = 3$, once again showing an inverse correlation between the z_{mm} and z_{mc} numbers.

In conclusion, we see that a major feature of the low-carbon carbides, where—unlike the high-carbon ones—a high density of C-cells is not required, is that low density C-cell structures can be formed in which the cells are fitted together in arrangements whose complexity requires much intermediate material but which also enables more than six M atoms to group round each C atom. Only two of the three types of C-cell allow this, the trigonal prism and the square antiprism, and these form the basis of the complex carbide structures.

REFERENCES

D.J. DYSON and K.W. ANDREWS, *J. Iron Steel Inst.*, (1969) **207**, 208.

E.J. FASISKA and G.A. JEFFREY, *Acta Cryst.*, (1965) **19**, 463.

F.C. FRANK and J.S. KASPAR, *Acta Cryst.*, (1958) **11**, 184.

G. HÄGG, *Zeit. Phys. Chem.*, (1931) **12**, (B), 33.

H.K. YAKEL, *Int. Metals Rev.*, (1985) **30**, 17.

2

THE AFFINITY OF TRANSITION METALS FOR CARBON

2.1 THE IMPORTANCE OF THE d-SHELL

TO ADVANCE beyond the simple sphere-packing model of the previous chapter it is of course necessary to consider the chemical interactions between transition metal and carbon atoms. The outstanding ability of transition metals, particularly the earlier members of the long periods, to combine strongly with carbon to form hard, refractory, carbides stems from their unique feature: the partly-filled d states in the valency shells of their atoms. The corresponding energy bands of d character, formed when many such atoms come together as a metal, remain partly filled; i.e. the energies of the d states in the metal straddle the Fermi level.

The valency shell of a neutral carbon atom is also partly filled, with 4 electrons shared among its 2s and 2p states, the total capacity of which is 8 electrons. When such an atom is put into a metal there is strong electrostatic resistance to its changing its complement of electrons appreciably. Thus the valence states of the carbon atoms remain partly filled and so also straddle the Fermi level of the metal. It follows that in transition metal carbides the metal and carbon atoms both provide, at about the same energy, a high density of partly filled valency states. This is an ideal starting condition for a strong covalent interaction of the two sets of states, resulting in the formation of bonding hybrid states, a few electron-volts (eV) below the Fermi level, and antibonding hybrids in a corresponding energy range above the Fermi level. The capacities of these two hybridised bands are such that a majority of the valency electrons, from the metal and carbon atoms, go into the bonding states where their drop in energy represents the chemical affinity which provides the cohesion of these carbides.

The concept of covalent bonding involving the carbon atom raises an immediate question from the familiar tendency of this atom to form a tetrahedrally-directed quartet of sp^3 hybrid orbitals, as in diamond and numerous chemical compounds. Why then does, for example TiC, both atoms of which have four valence electrons, form in an NaCl-type crystal structure with $z_{cm} = 6$, rather than in a ZnS-type where the $z_{cm} = 4$ would

11

allow tetrahedral bonding? The answer is in the electronic structure of the titanium atom, with five 3d orbitals and one 4s, that provides opportunity for six covalent bonds. It is a general understanding in quantum chemistry that the most stable covalent structures are those in which all of the stable orbitals of each atom are used in bonding formation.

While the above gives an understanding of the affinity of at least some transition metals for carbon, there remains the question of why the association of these elements, for example in TiC, is preferred to that of the uncombined components, titanium metal and graphite, to the extent of releasing an energy, $\Delta H = 1.91$ eV per TiC unit. In fact, it is difficult to give a convincing qualitative reason for this because, although 1.91 is a fairly large heat of formation, it represents only a small difference between the cohesive energies of the carbide and its constituents. Thus, relative to the energy levels of the free neutral atoms in their ground states, the cohesive energies, U, of titanium and graphite are -4.85 and -7.38 eV, respectively, so that $U = -12.23$ for the components and -14.14, per TiC unit, for the carbide; a difference of less than 16 percent, which sets a difficult challenge to qualitative understanding. In fact, the position is worse than this because, in forming the various solids, the electronic structures of the free atoms have to be promoted from their ground states into higher energy states which are suitable for the bonding interactions. In the ground state of the titanium atom, two of its valence electrons are in 4s states and two in 3d ones, i.e. the configuration is s^2d^2; but it is generally accepted (cf Pettifor, 1983) that in the metal the preferred state is sd^3, for which the promotion energy is $U_{prom} = 1.9$ (e.g. Gelatt et al., 1977). For carbon the promotion energy, from the s^2p^2 ground state, is both large and uncertain. For the promotion of one of its two 2s electrons to a 2p state, the value 7.1 eV has been suggested (van Vleck, 1933). If this same value were to apply in the carbide, then the cohesive energy, U_p relative to the promoted states of the free atoms is $-U_t = (4.85 + 1.9) + (7.38 + 7.1) = 21.23$ for the constituents; and 23.14 for the carbide, a difference of only 9 percent. Such differences cannot be convincingly explained by purely qualitative arguments and in fact, even quantitative ones based on sophisticated theory, described later, can explain them only approximately.

2.2 TRENDS ACROSS THE LONG PERIODS

The affinity for carbon is strongly evident among the elements at the beginning of the long periods, especially Ti, Zr and Hf, but it diminishes in

Table 2.1 Heats of formation, ΔH, of transition metal carbides, at room temperature, from the metals and graphite in standard states, in electron–volts per carbon atom. Italics indicate hypothetical carbides. Bracketed values are estimated from other thermodynamical information, in some cases by extrapolation.

Carbide	M sublatt.	ΔH	Carbide	M sublatt.	ΔH	Carbide	M sublatt.	ΔH
ScC	FCC	1.35	*YC*	FCC	0.95	*LaC*	FCC	(−0.45)
TiC	FCC	1.91	ZrC	FCC	2.04	HfC	FCC	2.17
VC	FCC	1.06	NbC	FCC	1.46	TaC	FCC	1.48
V_2C	CPH	1.43	Nb_2C	CPH	2.02	Ta_2C	CPH	2.16
CrC	FCC	(−0.01)	MoC	simple hex.	0.13	WC	simple hex.	0.42
Cr_3C_2	compl.	0.442	Mo_2C	CPH	0.48	W_2C	CPH	0.54
Cr_7C_3	compl.	0.555						
$Cr_{23}C_6$	compl.	0.567						
MnC	FCC	(−0.13)	*TcC*	FCC	(−0.66)	*ReC*	FCC	(−0.56)
Mn_7C_3	compl.	0.13						
Mn_3C	compl.	(0.42)						
$Mn_{23}C_6$	compl.	(0.53)						
FeC	FCC	(−0.43)	*RuC*	FCC	(−0.76)	*OsC*	FCC	(−0.92)
Fe_3C	compl.	(−0.22)						
CoC	FCC	(−0.53)	*RhC*	FCC	(−0.74)	*IrC*	FCC	(−0.71)
Co_3C	compl.	(−0.22)						
NiC	FCC	(−0.64)	*PdC*	FCC	(−0.66)	*PtC*	FCC	(−0.64)
Ni_3C	compl.	(−0.30)						

those further along these rows of the periodic table, so much so in the second and third long periods that no carbides are formed by the metals following Mo and W. These trends are shown by the heats of carbide formation, i.e. the energy released, in electron-volts (eV) per carbon atom, at room temperature, as a result of the reaction,

$$yM + C \rightarrow MyC. \tag{2.1}$$

Values are given in Table 2.1. The unbracketed ones are from standard tabulations of measurements. The others are estimated from thermodynamical information, in some cases by extrapolation (Häglund et al., 1991;

Guillermet et al., 1992, 1993). We see that ΔH drops, from about 2 per carbon atom, in group IV, steadily group by group, to near zero in group VII and thereafter.

The trends have been explained in terms of the increased filling of the d states in the later members of each period. (Gelatt et al., 1983; Häglund et al., 1993). The basis of this goes back to elementary quantum chemistry and the distinction between *bonding* and *antibonding* orbitals. It will be recalled that when, e.g., two hydrogen atoms, each with one electron in its 1s atomic orbital, are brought together they no longer have separate atomic orbitals. Instead, they behave as a single system in which the two nuclei share communal orbitals. Since each nucleus, by its proximity to the other atom, brings some attractive electrostatic potential into the nearby regions of the electron cloud of this other atom, it might be thought that the electron of this atom would shift slightly, into this region between the atoms, to take advantage of this. This indeed happens when the communal orbital is of the bonding type, but there is an equally significant alternative configuration.

Near each nucleus, the communal orbital closely resembles the original, unperturbed, atomic orbital of that nucleus. The key point is that the wave function of this atomic orbital has a choice of signs, positive or negative. This is of no physical consequence for a single atom, because its electronic distribution depends on the square of the amplitude of the wave function and so is independent of its sign. But it is important when two such atoms come together, because the communal orbitals reduce at the two nuclei to atomic wave functions which have either the same sign or opposite signs. If the signs are the same, then the situation leading to the bonding orbital, outlined above, holds. But when the signs are opposite the amplitude of the communal orbital must change from positive to negative at all points on a surface between the two nuclei. In other words, this orbital has a *nodal surface* between the nuclei, on which its amplitude is zero. In a simple case such as that of the two hydrogen atoms this is a plane, symmetrically bisecting the line joining the nuclei. The presence of a node means that the electron density is zero there, i.e. in this case the electrons avoid the region between the nuclei. The communal orbital is then of the *antibonding* type.

An electron in a bonding orbital has lower energy, i.e. is more stable than it was in its unperturbed parent atom, since it is able, in the region between the nuclei, to enjoy the attractive electrostatic potential there from the other atom. It thus provides a bond between the atoms. By contrast, an electron in an antibonding orbital, because it has to avoid the nodal surface, is driven away from this attractive region, towards the other side of its atom, so that its energy level is raised. It thus provides a repulsion, an 'antibond', between the atoms.

Whether the system formed by two interacting atoms exists as a stable molecule or not then depends on the *bond order*, i.e. on the number of electrons in bonding orbitals minus the number in anti-bonding ones. Each orbital, being the spatial part of a quantum state, can hold two electrons of opposite spins. The hydrogen molecular ion, H_2^+, is stable because its single electron can go into the bonding orbital, i.e. bond order = 1. Similarly, the neutral hydrogen molecule, H_2, is more stable when its two electrons, with opposite spins, go into the bonding orbital; bond order = 2. This orbital is now full, so that if a third electron is introduced, as in H_2^- or He_2^+, the bond order is reduced, $2 - 1 = 1$, and the bond is weakened. Finally, the attempt to provide four electrons, as in He_2, fills both bonding and antibonding orbitals, so that the bond order = 0 and no stable molecule is formed.

These principles extend to the solid state except that, with many atoms involved, a great number of different communal orbitals are formed, running through the whole system, in some places with nodes between atoms, in other places without. As a result, these orbitals provide many different energy levels, depending on their numbers of nodes, lying in an energy band between the extremes of the lowest level, with no nodes, and the highest with nodes between all atoms.

The d band, so formed in a transition metal, can thereby be divided approximately into a lower half of bonding orbitals and an upper one of antibonding ones. This simple scheme, which implies a maximum bond order at the half-filled band, i.e. at Cr, Mo, W, and falls away symmetrically on either side of this, is roughly in accord with the distribution of observed cohesive energies in the transition metals.

In a carbide, say MC, the combination of the four s, p states contributed by the C atom with the five d states of the M atom would again produce a band, consisting approximately of lower and upper halves of bonding and antibonding states, respectively. While this picture could correctly indicate the reduced affinity for carbon of the later members of each period of transition metals, in which the nearly full band implies some occupancy of antibonding orbitals and hence a small bond order, it cannot explain the entire trend since it implies that the maximum bond order and cohesion should occur in the carbides of Cr, Mo and W, whereas in practice these carbides have much lower heats of formation than the preceding ones (Table 2.1).

Several additional effects have been considered (e.g. Gelatt et al., 1983;

Häglund et al., 1993). The proposal below is that the high density of valence electrons in these carbides increases the electron-electron repulsion, particularly when the later transition metals are involved. This view stems from the picture to be developed later (e.g. chapters 5 and 6) of a typical carbide, e.g. from a group VA metal, such as VC with NaCl-type structure. In this, each carbon atom claims four valence electrons from its metal neighbours to form, with its own four valence electrons, a set of saturated covalent bonds. Each vanadium atom then has one remaining d electron which it can use to form, with those from its twelve metal atom neighbours, dd covalent bonds. Compared with the pure metal the metal-metal bond strength is thus reduced fivefold, but the effect of this is of course more than compensated by the strong metal-carbon bonding.

Group IV A carbides, e.g. TiC, behave similarly except that in these there are virtually no d electrons left over after forming the metal-carbon bonds, so that the dd orbitals are practically empty.

It may be assumed for simplicity, on the basis of generic expressions (Harrison, 1980) for bond interactions between orbitals on neighbouring atoms (particularly pd between the carbon 2p states and the metal d ones; and dd between metal atoms) that the basic interactions are intrinsically similar in all these cases but modified in several ways. The two most obvious modifications are due to different interatomic spacings and in the number of remanent dd bond electrons in the carbide.

If these only were active then, apart from the effect of differences in spacing, the heat of formation of VC would be expected to be as large as that of TiC. But there is also the effect of valence electron concentration as reflected in the presence of one electron per metal atom in dd orbitals in VC, but not in TiC. Consider, from this point of view, vanadium metal which is gradually filled with carbon to become VC. At first the metal atoms are bonded with five electrons per atom in dd orbitals. As the carbon content is increased these orbitals are weakened by the increasing removal of electrons to the carbon–metal bonds. The same of course occurs with titanium. But the difference is that the electrons in the carbon–metal orbitals and those in the dd ones interact repulsively and so raise their energy levels. Although the number of electrons in the carbon–metal orbitals is the same in VC and TiC, there are (virtually) none in the dd ones of TiC, but some in the VC ones and more in carbides of later transition metals such as chromium and manganese. The repulsive interaction will thus be stronger in these and so diminish the heat of formation. This will be discussed further in Chapter 8.

2.3 QUANTITATIVE METHODS

One of the features of modern electron theory is the greatly improved precision of its calculations of band structures and cohesive properties of various crystalline solids including transition metal carbides, even though the most accurate of these, based on *local density-functional* (LDF) methods, do not usually give a simple qualitative picture of the processes at work in the materials. Much of the credit for the improvement belongs to the powerful modern computer, the resources of which are fully stretched in these calculations, but a major part has been played by better methods of approximation stemming from the LDF theory.

The central task is the impossible one of solving Schrödinger's equation for an enormous number of nuclei and electrons, all in motion and mutual interaction. Even in the most ambitious *ab initio* calculations, in which the only input data are the nuclear charges of the constituents and the crystal structure of the solid, numerous approximations have to be made to reduce this to a tractable problem. An obvious one is to regard the heavy, slow-moving, nuclei as fixed in position when calculating the distribution of electrons round them (Born–Oppenheimer approximation). Another, sometimes made for many-electron atoms, is to regard the inner electrons in filled shells as unperturbed by what goes on outside the atom. In this 'frozen core' approximation these electrons are thus regarded as remaining in their free-atom quantum states and energy levels and so do not have to be considered explicitly in the Schrödinger equation. Their effect on the valency electrons is taken into account by regarding them as changing the actual electrostatic potential of the nuclei into a 'pseudopotential' which can be calculated by standard methods.

There remains the difficult problem: how to handle the many-electron Schrödinger equation for the valence electrons. The classical and essential first step here is to convert the many-electron equation into a set of soluble one-electron equations, one for each electron considered, by the Hartree approximation. It is assumed in this that each electron moves in a time-averaged field of the cores and all other valence electrons. The 'one-electron' equation for this electron is solved to give a wave function for it, from which the density distribution of this electron in the system is deduced and fed back into the Schrödinger equation as part of the time-averaged field of some other electron. The equation is solved again, for this other electron, and the ensuing process repeated, again and again. Such iterations are continued until the successive solutions no longer show significant differences. This is the self-consistent Hartree method for reducing the problem to tractable one-electron equations.

Although an essential method for all such many-electron problems, its weakness is that it deals only with time-averaged interactions of the electrons, not with the actual 'instantaneous' ones. There are two of these between any given pair of electrons. First, if they have the same spins they avoid each other by the Pauli exclusion principle, an effect usually described as an exchange interaction. Second, irrespective of their spins all electrons repel one another electrostatically, leading to a mutual avoidance known as the correlation effect. The general result is that each electron carries round with itself an *exchange-correlation hole*, a kind of flexible bubble in which no other electrons are likely to be found, the size of the bubble being such that the net positive charge in it balances the negative charge of the electron.

The method in LDF theory is to represent the exchange-correlation effect by a potential, V_{xc}, which is added to the Hartree potential or pseudo-potential. In this way, the Schrödinger equation is reduced to a set of one-electron equations, one for the wave function ψ_i of each electron i, at position r, i.e.

$$\left[-\frac{\hbar^2}{2m}\nabla^2 + V_{ext}(r) + V_H(r) + V_{xc}(r)\right]\psi_i(r) = \epsilon_i\psi_i(r), \qquad (2.2)$$

where

$$\hbar = h/2\pi \ (h = \text{Planck's constant}),$$

$$m = \text{mass of electron},$$

$$\nabla^2 = \frac{\partial^2}{\partial x^2} + \frac{\partial^2}{\partial y^2} + \frac{\partial^2}{\partial z^2},$$

$$V_{ext} = \text{the field from the nuclei},$$

$$V_H = \text{the Hartree (time-averaged) field from the other electrons},$$

$$\epsilon_i = \text{the energy level of the electron } i.$$

The LDF theory provides an exact expression for V_{xc} for the ground state of the system, but to evaluate it we would need to know the density of electrons, $n(r)$, at all points, r, in the system (which is implied by the use of the term *functional*) and since $n(r)$ can be known only when all the solutions $\psi_i(r)$ have been found, the theory, being thus non-local, is still fairly impractical at this stage. However, a good approximation can be made by the *local density assumption* (LDA) which, for the purposes of evaluating V_{xc}, involves approximating the actual electron density distribution, $n(r)$, by a uniform density equal to the local value at the point, r. The non-locality then disappears from eqn. 2.2, which thus becomes solvable by the self-consistency method. In this one starts with a guessed density $n_0(r)$, usually found by superimposing free atom electron clouds, from which the V terms are deduced. The one-electron equations are then solved to give the $\psi_i(r)$, from which a revised density distribution $n_1(r)$ is obtained and used to repeat the cycle of calculations, to give an improved $n_2(r)$. The self-consistent solution is reached when further iterations produce insignificant further improvements.

In evaluating the total electronic energy of the system from the energy levels, ϵ_i, given by the solution, there is a technical point to be noted. The value of ϵ_i obtained for electron i includes the energy of its interaction with all other electrons, among which for example is electron j. When the corresponding value of ϵ_j is found, this same interaction between i and j is included again, so that the sum of all the electron energies ϵ_i involves double-counting these interactions. A subtraction has thus to be made of one set of them to give the total energy correctly.

This basic LDF theory can be improved in many ways. For example:

(1) More accurate results can be obtained if the total population $n(r)$ of electrons is regarded as two distinct species, with up or down spins, which are each evaluated separately in the calculations. This is the *local spin density*, (LSD), or *spin-polarized* version of the theory.

(2) When an electron is in a very inhomogeneous region of the system, for example near a free surface, its exchange-correlation hole is much distorted, e.g. pulled to one side. Such effects can be allowed for by *generalized gradient approximations* (GGA).

(3) Particularly in the heavier elements (the 5 d series) electron velocities can be high enough to be relativistic. Refinements to take this into account are often made.

Having thus overcome the many-electron problem, in ab initio theory, there remains the many-atom problem to be tackled. Two general methods exist for this: (i) cluster approximations (ii) the perfect crystal approximation. In cluster methods only a small sample of the system is evaluated, for example a cage of nearest neighbours round a central atom, small enough to be handled in toto in the computer as a kind of 'molecule' embedded in a given environment. The advantage of these methods is that irregular configurations of the atoms can be studied; a disadvantage stems from the somewhat cavalier treatment of the role of the environment.

The alternative assumption of a perfect crystal structure allows one to use the translational symmetry to describe the entire, macroscopic, system in terms of what goes on in a single crystal cell. Most importantly it enables the *Bloch theorem* to be used to define the general form of all wave functions in the system, i.e. as

$$\psi(r) = u(r)e^{ikr}, \tag{2.3}$$

which represents a set of atomic-like functions $u(r)$, the same at each crystal site except for a change in phase from site to site which is indicated by the sinusoidal term, $\exp(ikr)$, with *wave number* k, where $|k| = 2\pi/$wavelength.

The most common method of representing the crystal potential, i.e. the field in which the electrons move, in this case is a development of the classical Wigner–Seitz method known as a *muffin-tin potential* (Ziman, 1964), named from its resemblance to the cooking tin. An imaginary sphere is constructed round each atomic site in the crystal and inside it the potential is assumed to be spherically symmetrical; rather like the potential of a free atom but with a different boundary condition at the surface of the sphere. In the interstitial regions between the spheres the potential is assumed to be constant. These potentials are evaluated by the self-consistent LDF or LSD methods. Wave functions are constructed in the two regions. In the interstices they are simply free electron functions, i.e. $u(r) =$ constant there, and inside the sphere the spherical potential gives them an atomic-like form. The parameters in these functions are adjusted so that they match smoothly at the surface of the sphere.

In this way, for each value of k a crystallographically repeating function $u(r)$ and hence a $\psi(r)$ is found. Various values of k have to be considered, so spreading the energy levels, ϵ, of the corresponding $\psi(r)$ into a band, but the crystal symmetry confines these k values to a Brillouin zone. Further crystal symmetries, e.g. rotation and reflection, then cause the ϵ,k relations to repeat in various parts of the Brillouin zone. A minimum region (non-repetitive) of the zone is then selected and ϵ evaluated for a selection of k points in it. In this way an ϵ,k band structure is deduced from which the density of states and subsequent properties can be evaluated.

Several variants of these methods have been applied to the electronic structures of carbides. The main ones are:

LAPW   (linear augmented plane wave) in which plane waves in the interstitial regions of the potential are matched to atomic-like radial solutions inside the spheres.

LMTO   (linear muffin tin orbitals) in which free atom orbitals, inside the spheres, are replaced by orbitals deduced directly for the spheres.

KKR   (Korringa, Kohn and Rostoker) in which the wave function is deduced by scattering of waves off the spheres.

ASW   (augmented spherical wave), a variant of the LMTO method.

FLAPW   (full-potential linear augmented plane wave) which is a modification of LAPW made more accurate by using a correct potential in the interstitial regions in place of the assumed constant one.

FLMTO   (full-potential linear muffin tin orbital), a similar modification of LMTO.

A good summary of the application of these and other methods to the electronic structure of carbides (and nitrides) has been given by Gubanov et al. (1994). Recent elementary introductions to the underlying theory have been given by Cottrell (1988), in the book edited by Pettifor and Cottrell (1992), and by Sutton (1993).

2.4 DENSITY OF STATES IN CARBIDES

The many studies of transition metal carbides made by these methods—and others, considered in the next chapter—have mainly produced consistent results (Gubanov et al., 1994). Figure 2.1, which is typical, shows the density of states distribution for stoichiometric TiC, calculated by Neckel et al., (1976) using the LCAO method described later (Chapter 3).

In this the s, p and d bands consist mainly of well-defined peaks, as expected from the covalent character of the carbon bonding and the intrinsic narrowness of d bands in transition metals. Not shown is a second s band, representing free electron states formed from the 4s states of titanium, which has been pushed to high energy levels by repulsion from the carbon 2s states. The relative positions of the peaks along the energy scale, in relation to the Fermi level E_F, are significant. The s band, which mainly represents the 2s states of the carbon atom, lies at about 10 eV below E_F, and about 6 eV below the lowest peaks of the p band. This shows that the carbon does not

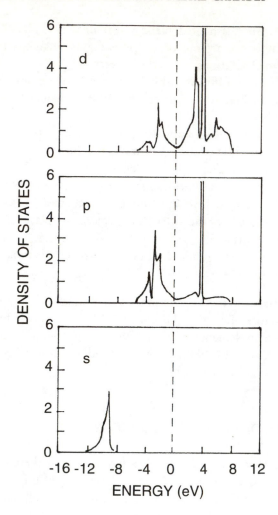

Fig. 2.1 Densities of s,p,d states in TiC, calculated by Neckel et al. (1976) using the LCAO method. The energy, in electron-volts, is relative to the Fermi level. The density of states is in arbitrary units.

form the sp^3 hybrids characteristic of carbon compounds, since if it did these two bands would fall in the same energy range, i.e. that of sp^3 hybrid orbitals. The same argument reveals the pd hybridization between the carbon 2p and titanium 3d orbitals, as indicated by the coincidence of the two sets of peaks, one centred at about -3 eV, the other at about $+4$ eV. These are clearly the bonding and antibonding hybrids, respectively, between these orbitals; and the band split, about 7 eV, indicates the strength of this interaction between the carbon and titanium atoms.

It will also be seen that there are some additional d states, at around +2 and +6 eV, which are much less well matched by corresponding p states. These are interpreted as bonding and antibonding dd hybrids, the orbitals of which are less suitably oriented for interaction with the carbon orbitals. Further investigation has shown that these d states have the symmetry known as t_{2g} in a cubic crystal, with their lobes lying along < 1 1 0 > directions, in contrast to those of e_g symmetry, the lobes of which point along < 1 0 0 >, straight at the carbon atoms.

The position of the Fermi level in relation to these bands shows that the s band is completely full (2 electrons per C atom, i.e. per TiC unit), the bonding pd bands are also completely full (6 electrons per TiC unit since there are 3 p states per C atom available for this hybridization), and the antibonding pd bands are completely empty, as are also the bonding and antibonding dd hybrids. The latter indicates an interesting feature. The dd bonding which makes pure titanium a strong metal has virtually disappeared in TiC. Practically all the valence electrons of the Ti atoms have been transferred to pd bonds with the carbon atoms. The other significant point is that all the valence electrons, 8 per TiC unit, are accommodated in low-lying bonding orbitals, with none in antibonding ones, so favouring strong cohesion.

Corresponding calculations, made for the stoichiometric NaCl-type carbides of the other group IV and V metals (Neckel et al., 1976; Häglund et al., 1993; Gubanov et al., 1994), show a striking similarity in their density of states distributions, so that to a fair approximation they can all be discussed in terms of a common rigid band which is filled to various levels by the different numbers of valence electrons available. Fig. 2.2 shows this and indicates the Fermi levels for 7(Sc,Y,La), 8(Ti,Zr,Hf), 9(V,Nb,Ta) and 10(Cr,Mo,W) valence electrons per MC cell. We see that for Group V metals the Fermi level is raised into the region of the dd bonding orbitals, with one electron per cell in these. No carbides of this type are formed from group VI metals, which is consistent with the elevation of the Fermi level, at 10 electrons per cell, into the region of the antibonding pd states, although Mo and W form non-stoichiometric NaCl-type carbides at about $MC_{0.6}$.

Band calculations have also been made for carbides with other structures (Gubanov et al., 1994). Fig. 2.3 shows two examples, WC with a simple hexagonal metal sublattice (Mattheiss and Hamann, 1984; Price and Cooper, 1989); and Fe_3C (Häglund et al., 1991). Both density of states

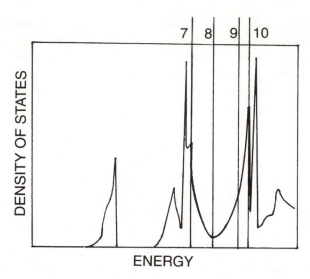

Fig. 2.2 Total density of states distribution (schematic) for stoichiometric NaCl-type carbides. The Fermi levels for various numbers of valence electrons per MC unit are indicated.

distributions are similar to those of the other carbides, with a low-lying s band and with the close correlation between p and d states indicating a filled band of bonding pd hybrids down to about -8 eV below the Fermi level.

In the case of WC, although there is pd correlation in the range 0 to -2 eV, indicating some pd hybridization there, the d band is much larger than the p one, so that some of its states are not involved in this hybridization. In fact they approximately match another d peak at about $+2$ to 3 eV, which suggests that these two d peaks are (apart from the small pd hybridization component of the lower one) bonding (filled) and antibonding (empty) dd hybrids between metal atoms.

In the case of Fe_3C, because the iron atoms have more d electrons and because there are fewer carbon atoms to take some of them into pd bonds, a much larger fraction of d electrons is available for dd bonding between the metal atoms. This accounts for the large range of high d state density from -3 to $+1$ eV, with little matching p density. The complexity of the band structure in this range, with many small peaks is a consequence of the complexity of the Fe_3C crystal structure (Fig. 1.6).

The formation of d peaks representing bonding dd hybrids (filled) and

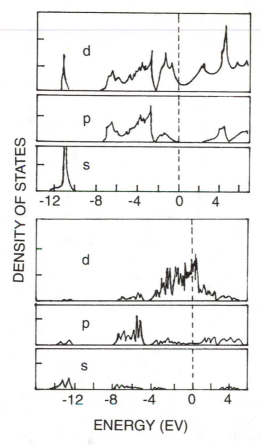

Fig. 2.3 Densities of s,p,d states. Upper diagram: hexagonal WC (after Price and Cooper, 1989). Lower diagram: Fe_3C, assumed non-magnetic, after Häglund et al., 1991). The positions of the Fermi level are indicated. The density of states is in arbitrary units.

antibonding ones (empty), below and above the Fermi level respectively, as a consequence of the inability of the carbon atoms in these carbides to take up all the d electrons into pd bonds, is matched by a similar effect in the band structure of non-stoichiometric titanium carbide, as calculated by Redinger et al. (1985, 1986) using an augmented plane wave method. They considered $TiC_{0.75}$ in which the one vacant C-site in four was arranged in a (hypothetical) ordered structure, i.e. with the central site in Fig. 1.2a empty. Fig. 2.4 shows the total density of states. As well as the usual formation of a low-lying s band and the pd hybrids which peak at about 3 eV below the

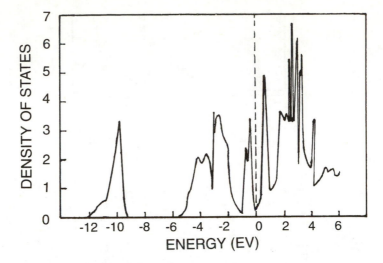

Fig. 2.4 Total density of states distribution for ordered $TiC_{0.75}$, after Redinger et al. (1985). The energy, in electron-volts, is relative to the Fermi level. The density of states is in arbitrary units.

Fermi level (with their antibonding peak at about 3 eV above it), there are, compared with TiC, two extra peaks about 1.2 eV apart which straddle the Fermi energy. These are the bonding and antibonding hybrids representing dd interactions between the metal atoms. The lower peak contains those dd electrons surplus to the pd bonds.

Despite great differences of crystal structure among these various carbides there is much similarity in their density of states distributions when the limited capacity of the carbon, to remove d electrons from metal-metal bonds to carbon-metal bonds, is taken into account.

2.5 COHESIVE ENERGIES

From the total density of states, $n(\epsilon)$, at the various electronic energy levels, ϵ, the electronic energy of the system (at $0°K$) can be found by summing the electronic energies, i.e.

$$E = 2\int \epsilon\, n(\epsilon)\, d\epsilon - DC, \qquad (2.4)$$

where DC is the correction for double counting (cf section 2.3). The integral is over all occupied states up to the Fermi level and the 2 represents the two electrons of opposite spins which fill a state. As noted by Harrison (1980), even when the density of states distribution consists, as above, of numerous peaks and valleys, this energy integral rises fairly uniformly as its upper limit is increased towards the Fermi level. This justifies Friedel's (1969) simple but successful modelling of the d band in transition metals by a rectilinear density of states without peaks and valleys.

By subtracting from the above E the corresponding values of E for the constituent elements, evaluated for consistency by the same method, the heat of formation of the compound is obtained. Several investigators have calculated theoretical values of ΔH in this way (Gellatt et al., 1983; Price and Cooper, 1989; Häglund et al., 1991, 1993). For carbides such as TiC they all obtained a positive ΔH, indicating a large affinity, but their values are generally higher than the observed ones, although with the most accurate methods, e.g. the FLMTO method of Price and Cooper, the calculated total energy of TiC is only about 26 percent too large. It appears that the discrepancy is mainly due to the use of the local density assumption (LDA), or the local spin density one (LSD), for the exchange-correlation potential V_{α} (cf eqn. 2.2). Although the absolute value of the cohesion is thus given only inaccurately, the theory provides a good indication of the trend in cohesion across the transition metal series. For example, Häglund et al. (1991) have shown that the variation in theoretical cohesive energies calculated by the LMTO method, from ScC to NiC in the 3d series, almost perfectly matches that of the values deduced from observational data.

REFERENCES

A.H. COTTRELL, *Introduction to the Modern Theory of Metals*, The Institute of Metals, London (1988).

J. FRIEDEL, in *The Physics of Metals, I—Electrons* (ed. J.M. Ziman), Cambridge University Press (1969).

C.D. GELATT, H. EHRENREICH and R.E. WATSON, *Phys. Rev. B*, (1977) **15**, 1613.

C.D. GELATT, A.R. WILLIAMS and V.L. MORUZZI, *Phys. Rev. B*, (1983) **27**, 2005.

V.A. GUBANOV, A.L. IVANOVSKY and V.P. ZHUKOV, *Electronic Structure of Refractory Carbides and Nitrides*, Cambridge University Press (1994).

A.F. GUILLERMET, J. HÄGLUND and G. GRIMVALL, *Phys. Rev. B*, (1992) **45**, 11557; (1993) **48**, 11673.

J. HÄGLUND, G. GRIMVALL, T. JARLBORG and A.F. GUILLERMET, *Phys. Rev. B*, (1991) **43**, 14400.

J. HÄGLUND, A.F. GUILLERMET, G. GRIMVALL and M. KÖRLING, *Phys. Rev. B*, (1993) **48**, 11685.

W.A. HARRISON, *Electronic Structure and the Properties of Solids*, Freeman, San Francisco, (1980).

L.F. MATTHEISS and D.R. HAMANN, *Phys. Rev. B*, (1984) **30**, 1731.

A. NECKEL, P. RASTL, R. EIBLER, P. WEINBERGER and K. SCHWARZ, *J. Phys. C: Sol. State Phys.*, (1976) **9**, 579.

D.G. PETTIFOR, in *Physical Metallurgy* (eds. R.W. Cahn and P. Haasen), Elsevier, Amsterdam (1983).

D.G. PETTIFOR and A.H. COTTRELL (eds), *Electron Theory in Alloy Design*, The Institute of Materials, London (1992).

D.L. PRICE and B.R. COOPER, *Phys. Rev. B*, (1989) **39**, 4945.

J. REDINGER, R. EIBLER, P. HERZIG, A. NECKEL, R. PODLOUCKY and E. WIMMER, *J. Phys. Chem. Solids*, (1985) **46**, 383; (1986) **47**, 387.

A.P. SUTTON, *Electronic Structure of Materials*, Clarendon Press, Oxford (1993).

J.H. VAN VLECK, *J. Chem. Phys.*, (1993) **1**, 177, 219.
J.M. ZIMAN, *Principles of the Theory of Solids*, Cambridge University Press (1964).

3

BONDS

3.1 FROM BANDS TO BONDS

THE *ab initio* methods of the previous chapter can provide quantitative results without help from empirical data but they are also rather abstract and do not suggest intuitive pictures of the bonding between atoms. The data streams which their computer calculations produce need interpretation at a different level of theory based on simpler and more pictorial concepts, in much the same way that Pauling extracted from general quantum mechanics chemically useful ideas such as resonating valence bonds. In addition, the heavy computation required by the *ab initio* methods, even for the simplest crystal structures, makes them impracticable for complex or irregular structures and lattice defects.

There is thus a need for an alternative approach, if only to supplement the *ab initio* one; one which, buttressed where necessary by empirical data or values from the *ab initio* methods, can be visualized qualitatively and is readily adaptable to complex structures. This clearly has to be a more 'chemical' approach, concerned directly with bonds between atoms, one that emphasizes the importance of the *local* environment of an atom for its cohesion. Such an approach will be sought in this chapter.

3.2 THE LCAO METHOD

The basic step is to represent the system with its communal orbitals as a layout of slightly overlapping, unperturbed, atomic orbitals. This is the *LCAO method* (linear combination of atomic orbitals) of quantum chemistry, also known in its application to crystalline solids as the *tight-binding method*. In the case of the H_2 molecule, for example, the two nuclei share communal orbitals, bonding and antibonding, but near each nucleus these orbitals closely resemble the atomic orbitals of the free atom. The LCAO method takes advantage of this by modelling the communal orbitals as a pair of slightly overlapping atomic orbitals. These may have the same sign and so represent a bonding hybrid; or opposite signs for an antibonding one. This approach is particularly suitable for the transition metals because the cohesion of these is mainly due to interaction between the valence d states and the electron clouds of the atomic d shells overlap only slightly at the interatomic spacings of these metals. It is also suited to the carbides, because here the valence electrons are substantially localized in C-M covalent bonds, with any electrons surplus to this correspondingly providing dd interactions between neighbouring metal atoms.

The LCAO method was in fact introduced in the earliest days of the electron theory of solids (1928), as the tight-binding method, by Bloch, who constructed wave functions for the solid from atomic orbitals as follows. Suppose that a given atom has a certain atomic orbital $\phi(r)$, where r is the (vectorial) distance from the nucleus. Form a solid from many such atoms, by putting atom i in the vector position r_i, so that its wave function at an arbitrary point r is $\phi(r - r_i)$. Then the Bloch tight-binding wave function, ψ, is the sum over all atoms of all such wave functions, each with a phase $exp(ikr)$ as in eqn. 2.3; i.e.

$$\psi_k(r) = \sum_i \phi(r - r_i) \, e^{ikr}. \qquad (3.1)$$

There is of course a separate Bloch function for each type of atomic orbital, e.g. d_{xy}, d_{yz}, d_{zx}, $d_{x^2-y^2}$, d_{z^2}, considered; and it is formed in each case by a sum of this type.

The next step is to set up a scheme for describing the overlaps between atomic orbitals. Because they practically vanish beyond about one atomic radius from their parent nucleus, only those between nearby atoms need be considered, which brings this theory of condensed matter close to that of molecular chemistry. There are various types of such overlaps between two neighbouring atoms, depending upon the particular atomic orbitals involved and the relative positions of the atoms. Fig. 3.1 shows a few examples identified by Blaha and Schwarz (1983) as particularly important in NaCl-type carbides. Here the 4-lobe orbitals represent 3d states on e.g. a titanium atom, oriented either along the cubic crystal axes (i.e. e_g symmetry) or at 45° to them (t_{2g} symmetry). The 2-lobe orbitals represent 2p states of carbon atoms. In two of the diagrams the orbitals meet 'head-on' (σ type), giving a large overlap and hence strong interaction. The third shows a 'sideways' (π type) overlap which is rather weaker. The examples in this Figure represent symmetrical situations where the orientations of the orbitals are aligned to that of the crystal structure. More generally, it is necessary along each interatomic axis to represent each state as a sum of components with

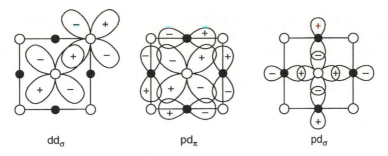

dd_σ $\quad\quad\quad$ pd_π $\quad\quad\quad$ pd_σ

Figure 3.1 Configurations of overlapping d and p orbitals in the (100) plane of an NaCl-type carbide (after Blaha and Schwarz, 1983).

different values of m (see below). Slater and Koster (1954) have given general expressions for this.

A significant feature of Fig. 3.1 is that all overlapping lobes have the same sign, so that these atomic orbitals model bonding orbitals. There are two comments about this. First, if the wave number k in the Bloch function is sufficiently altered, i.e. if these same diagrams are reconstructed for a distant point in the Brillouin zone, the phases of the atomic orbitals could reverse from one side of the cell to the other so that the overlaps would then become anti-bonding.

The second comment is that changes of sign restrict the types of overlap which produce interactions, as emphasized by Slater and Koster. Regard the vector from one atom to a neighbour as an axis of symmetry for the wave functions of the orbitals of these two atoms. A circuit taken round this axis will disclose this symmetry, as indicated by the angular momentum quantum number, m. This implies, round a circular circuit of path length L, an oscillation in sign of the wave function with a wavelength L/m. The allowed values of $\pm m$ are 0 for s states; 0, 1 for p; and 0, 1, 2 for d. If two atomic orbitals with different m values overlap, then in some regions they have the same sign but in equal amounts of other regions they have opposite signs so that their net interaction is zero; i.e. they are *non-bonding*. Only orbitals with the same value of m need be considered. The three types of overlap mainly of interest in the carbides are thus the dd_σ, pd_π and pd_σ of Fig. 3.1. There are several others, e.g. dd between two orbitals of $m = 2$ symmetry, but in general these can be disregarded, having weak interactions because of small overlaps or large difference in the atomic energy levels (e.g. as in the case of sd overlaps between carbon 2s and metal 3d orbitals).

Corresponding to each type of overlap such as $dd\sigma$ there is a corresponding overlap energy $V_{dd\sigma}$. In terms of these V, the interactions between two atoms can then be found, from the vector direction of one atom relative to the other, using the table of expressions for this given by Slater and Koster; and from the interatomic spacing, R, for which the relation $V \propto R^{-5}$ is often used (Andersen, 1973). Finally, the Bloch phase factor is introduced to give the dependence on position in the Brillouin zone.

The merit of this procedure is that it provides easier calculations of the energy of the states throughout the Brillouin zone. As a simple example, in an FCC crystal the interaction energy E_{xy} of a d_{xy} orbital on an atom, with the d_{xy} orbitals on its 12 nearest-neighbour atoms, varies along the z (cubic) axis of the Brillouin zone as

$$E_{xy} = 3V_{dd_\sigma} + 4V_{dd_\pi} \cos \Theta, \qquad (3.2)$$

where $\Theta = \frac{1}{2}k_z a$, with k_z as the wave vector of the Bloch function in this direction, and a is the FCC lattice constant. Thus Θ increases from $0°$ at the centre of the zone to $180°$ at its z boundary. Corresponding trigonometric expressions can be written for other combinations of orbitals and directions across the Brillouin zone. These are all simple analytical expressions by means of which one can without undue effort plot the pattern of E,k values in symmetrical directions across the Brillouin zone. The difficulty of course lies in obtaining values for V_{dd_σ}, V_{dd_π}, etc., for which there is no accurate *ab initio* theory, although Harrison (1980) has deduced some analytical expressions with approximate parameters. Nevertheless this LCAO theory is most useful as a labour-saving method used in conjunction with *ab initio* calculations. One then need make the heavy *ab initio* calculations at only a few places in the Brillouin zone, places where special symmetry simplifies them. The energies of those zone states deduced in this *ab initio* way can then be fitted to expressions such as eqn. 3.2, so determining the values of V_{dd_σ}, etc., from which the LCAO equations can finally be used to construct the Brillouin zone of E,k values and the density of states as in Fig. 2.1.

3.3 EFFECT OF COORDINATION NUMBER

The above representation of cohesion in terms of individual bonds between pairs of neighbouring atoms leads to a simple dependence of this cohesion on the number of neighbours, z, to a given atom in the solid. An obvious bond-counting approach suggests that it should be directly proportional, but this is not so when the coordination number z is greater than the number of bond orbitals. The expression to be used in such cases is of the approximate form

$$U_{bond} \simeq z^{\frac{1}{2}}h \qquad (3.3)$$

where h is the overlap integral or bond energy for one particular choice of bond partners in the coordination shell.

There are two approaches to this. That for tight-binding systems derives eqn. 3.3 by relating the covalent bandwidth to the coordination number, as outlined in section 3.4 below. The other, more qualitative but also more general, approach starts from Pauling's concept of the resonating valence bond (Heine and Hafner, 1989–90) and applies to all systems which can be represented by alternative patterns of covalent bonds, as for example in the classical case of the two Kekulé structures of benzene. The usual method for these is to construct the general wave function of the system as a weighted sum of individual wave functions for the alternative patterns of the bonds in the system. The energy is then expressed as a function of the weighting coefficients and the values of the latter deduced for the state of lowest energy. Provided that the frequency of resonance of the electronic structure through these alternative patterns is much higher than that of oscillation of the participating atoms, this general wave function with optimised weighting coefficients provides a good representation of the system. Normally such electronic frequencies are higher (but when they are not, as for example in a ferromagnetic state, the system exhibits broken symmetry).

It is a standard result of quantum mechanics that the electron energy for the state with the optimised sum of individual wave functions is lower than that of any individual contributing state. This follows from the uncertainty principle, since in the summed state the electrons have the additional freedom to explore all the alternative bond positions.

Eqn. 3.3 also emerges from more complete treatments. For example, a detailed analysis by Pettifor (1989) gives, for the leading term in the bond energy

$$U_{bond} = \tfrac{1}{2}\sum_{i}\sum_{j\neq i}h(R_{ij})\Theta_{ij} \qquad (3.4)$$

per state, per atom or molecular unit, with

$$\Theta_{ij} = 4\chi h(R_{ij})\, F\!\left(\frac{\rho_i + \rho_j}{2}\right), \qquad (3.5)$$

$$\rho_l = \sum_{k\neq l}h^2(R_{lk}), \qquad (3.6)$$

$$F(\rho) = \rho^{-\frac{1}{2}}, \qquad (3.7)$$

where the sum is over all atoms i in this molecular unit and over the (nearest) neighbours j of each atom i. The overlap integral h is a function of the interatomic spacing R_{ij}, as is also the bond order Θ_{ij}, which additionally depends on the other factors given above, with k summed over the nearest neighbours to atom l and where χ varies approximately parabolically with the degree of band filling up to about 0.42 for a half-filled band. Applied, for example, to a single atom in an FCC metal, with R_{ij}, R_{lk} and h all constant, eqn. 3.4 reduces to

$$U_{bond} = \sqrt{12}\,(2\chi h) \qquad\qquad (3.8)$$

in agreement with eqn. 3.3.

The bond order appears in eqn. 3.4 since, when the band is nearly empty of electrons there are few to resonate among the bonds; and when the band is nearly full there are few empty orbitals available to a resonating electron.

3.4 THE MOMENTS METHOD

An important advance in bonding theory, which in fact provided the formal basis for the $z^{\frac{1}{2}}$ effect above, was made by Cyrot-Lackmann (1967) and further developed by Friedel (1969) and Ducastelle and Cyrot-Lackmann (1970). Introductions to the theory have been given by Ducastelle (1991) and Sutton (1993). It involves describing the density of states distribution, $n(E)$ per atom, by its *moments*, as for a probability distribution. Taking E_0 as the mean energy of the band of states, the mth moment is

$$\mu_m = \int_{band} n(E)\,(E - E_0)^m\, dE. \qquad\qquad (3.9)$$

The advantage of this is that important features of the distribution are modelled by its first few moments. The zeroth moment, μ_0, is simply the integral of the distribution, i.e. the number of states per atom in the band. The first moment vanishes, $\mu_1 = 0$, from the definition of E_0. The second, μ_2, is the mean square deviation or variance of the distribution. The moments theory (e.g. Friedel, 1969, Cottrell, 1988, Sutton, 1993) shows that μ_2 for an atom in the solid is the sum of squares of its overlap interactions with its neighbours, i.e. ρ_l in eqn. 3.6. Thus when these all have the same h, it is

$$\mu_2 = z h^2. \qquad\qquad (3.10)$$

The root mean square deviation of the distribution is of the order of half of the band width W, so that

$$W \simeq 2z^{\frac{1}{2}}h. \tag{3.11}$$

For a partly (e.g. half) filled band in which the electrons predominantly occupy the low energy states the total cohesive energy increases proportionally with W. Thus eqn. 3.3 is confirmed.

A key feature of the moments theory, which greatly simplifies the application of LCAO bond theory to partly filled bands, stems from the relation of h to the hopping of valence electrons, to and fro, between overlapping atomic orbitals. For an electron to be considered in this way it must be regarded as localized initially in one of the atomic orbitals, say in ψ_1 on atom 1, and then later localized in ψ_2 on 2. But the two (stationary state) orbitals which span both atoms are $\psi_1 + \psi_2$ (bonding) and $\psi_1 - \psi_2$ (antibonding), suitably normalized. Hence to form an orbital for an electron localized in atom 1, the sum of these must be used, $\frac{1}{2}(\psi_1 + \psi_2) + \frac{1}{2}(\psi_1 - \psi_2)$, which of course is simply ψ_1 again. What this means is that, when the second atom is alongside, ψ_1 is no longer a stationary state. It has an uncertainty of energy equal to the difference between the bonding and antibonding energy levels, i.e. $\Delta E = 2h$, and so, from the uncertainty principle, has a limited lifetime, $\Delta t \simeq \hbar/\Delta E$ ($\hbar$ not to be confused with the above h). The time-dependent version of the theory shows that, after this order of time, the electron will be found in ψ_2 on atom 2; then after another Δt back on atom 1 again; and so on. The electron can thus be regarded as hopping to and fro between the two atoms at a frequency proportional to the overlap interaction energy, h. The second moment μ_2 of the local density of states at a given atom is thus the sum of all two-hop paths from that atom to its neighbours, multiplied by the squared interaction (i.e. hopping) energy for these jumps. Thus $\mu_2 = zh^2$ when these energies are all equal to h.

This interpretation of moments in terms of the number of hops which bring an electron back to its starting point, through the cyclic alternation of the signs of the combination of bonding and antibonding orbitals, from the uncertainty principle, greatly helps the extension of the theory to the description of the density of states using higher moments.

Eqn. 3.10 is a contracted version of the sum over all the neighbours j of a given atom i, i.e.

$$\mu_2 = \sum_j h_{ij} h_{ji}, \tag{3.13}$$

h_{ij} representing the hop from i to a particular atom j; and h_{ji} representing the return hop. If, as in the d shells of transition metal atoms, there is more than one kind of valence state on an atom, from and to which the hops can be made, then a separate $h_{ij} h_{ji}$ term exists for each of the various combinations of state-to-state hopping, and these have all to be included in the sum, eqn. 3.13.

3.5 HIGHER MOMENTS

While the second moment gives a useful first indication of a density of states distribution it is generally necessary to include higher moments to get a fuller picture of the band. The most important of these are the third and fourth, defined by

$$\mu_3 = \sum_{j,k} h_{ij} h_{jk} h_{ki},$$

(3.14)

$$\mu_4 = \sum_{j,k,l} h_{ij} h_{jk} h_{kl} h_{li},$$

(3.15)

for hopping from a given atom i.

The concept of the hopping of an electron or electrons between two different atoms leads naturally to $i \neq j \neq k \neq l$ in these expressions. It is necessary, however, to consider also the case of *on-site hopping*, which would for example include terms of the type $h_{ij} h_{jj} h_{ji}$ in eqn. 3.14. The meaning of h_{jj} is that it represents a 'hop' in which the electron remains in the same atom, j. In other words, during this 'hop', the electron stays in the valence electron energy level, ϵ_j, of this atom. We cannot simply put $h_{jj} = \epsilon_j$, however, because of the meaning of h, which measures the shift from the original valence level, ϵ_i, of the atom i brought about by the interaction with j. For example, when i and j overlap to form a bonding orbital between them at the energy level ϵ_b, then $h_{ij} = \epsilon_b - \epsilon_i$. On this basis, when evaluating moments for atom i, we have $h_{jj} = \epsilon_j - \epsilon_i$, and thus

$$h_{ij} h_{jj} h_{ji} = h_{ij} (\epsilon_j - \epsilon_i) h_{ji},$$

(3.16)

Such terms vanish, of course, when all the valence states are at the same energy level, which leads back to the case where $i \neq j \neq k \neq l$ again. But when, as in the carbides, different species of atoms are neighbours, with different values of ϵ_i and ϵ_j, they have to be taken into account.

The extra factors in μ_3 and μ_4 introduce several different hopping paths between the neighbouring atoms, some of which are shown in Fig. 3.2. The three-hop path starting from atom b is of course not possible in certain lattices such as simple cubic, but when the atomic energy levels differ a three-hop path such as that starting from atom c, including an on-site hop, can

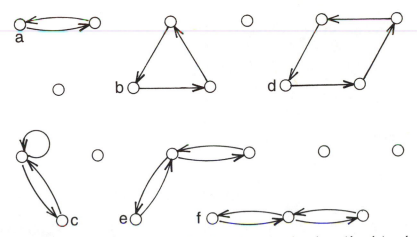

Figure 3.2 Examples of closed paths of two hops (*a*), three hops (*b* and *c*) and four hops (*d*, *e* and *f*), starting from atoms *a, b, c, d, e, f.*

occur irrespective of the lattice structure. Fig. 3.2 also shows three different types of four-hop paths. Of these the straight self-retracing path, from atom *f*, gives the strongest contribution to μ_4 (cf. Ducastelle, 1991). This is because changes of direction along a path, such as those from atoms *d* and *e* (other than diametric reversal, as in *f*), involve changing the mixture of ddσ, ddπ ... etc., components in the orbitals along these paths. Thus, if for example a ddσ component lies directly along one leg of the path, so favouring a strong bond, then along an inclined leg some or all of this ddσ component is replaced by ddπ or an even weaker one.

The moments μ_3 and μ_4 indicate differences in the shape of the density of states distribution. Since μ_3, from eqn. 3.9, is negative when the states for which $E < E_0$ are well strung out along the energy axis and those for which $E > E_0$ are densely bunched near E_0, then μ_3 indicates a *skew distribution* as shown in Fig. 3.3. Thus in an alloy or compound where the components have different atomic energy levels, as in eqn. 3.16, the component with the higher level (ϵ_i) and hence negative ($\epsilon_j - \epsilon_i$) in eqn. 3.16 tends to have a negative μ_3 and a corresponding skewness in its local density of states. By contrast, the other component then has a positive ($\epsilon_i - \epsilon_j$) and an inverse skew in its local density of states.

A small μ_4 indicates a *bimodal* distribution (Fig. 3.3) and is obtained from well-separated bonding and antibonding interactions. As μ_4 increases, the distribution gradually changes to a *unimodal* one. When $\mu_3 = 0$ these changes are conveniently measured by

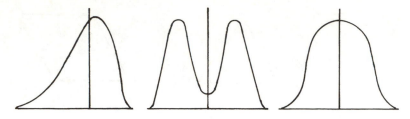

Figure 3.3 Skew, bimodal and unimodal distributions.

$$\beta = \frac{\mu_4}{\mu_2^2} \qquad (3.17)$$

equal to 3 for a gaussian distribution, 1.8 for a rectangular one and 1 for two sharp energy levels symmetrically above and below E_0. When $\mu_3 \neq 0$ the corresponding measure is

$$\beta = \frac{\mu_4 \mu_2 - \mu_3^2}{\mu_2^3} \qquad (3.18)$$

The moments theory has led to a remarkable deduction by Ducastelle and Cyrot-Lackmann (1971). Suppose that the cohesive energies of two alternative crystal structures, e.g. FCC and CPH, are being compared at various stages in the filling of the valence band, e.g. along the row of 3d transition metals. Consider the differences $\Delta\mu_m = \mu_m - \mu_m$ in their moments, where $m =$ 0, 1, 2, ... etc. Then, if $\Delta\mu_m = 0$ for all moments up to and including μ, the difference in electronic energy between the two structures must cross the zero line at least $m - 1$ times as the band is filled (in addition to zeros for the completely empty or full band). For example, because the FCC and CPH structures have the same nearest-neighbour distributions, $\Delta\mu_0 = \Delta\mu_1 = \Delta\mu_2$ $= \Delta\mu_3 = 0$ and so their energy difference must cross the zero line at least twice during the band filling. This confirms the results of direct calculations for FCC and CPH structures of transition metals (e.g. Pettifor, 1972) which show an alternation of stability as the d band is filled, with an energy difference never exceeding 0.05 eV per atom.

REFERENCES

O.K. ANDERSEN, *Sol. State Comm.*, (1973) **13**, 501, 511.

P. BLAHA and K. SCHWARZ, *Int. J. Quantum Chem.*, (1983) **23**, 1535.

A.H. COTTRELL, *Introduction to the Modern Theory of Metals*, The Institute of Metals, London (1988).

F. CYROT-LACKMANN, *Adv. Phys.*, (1967) **16**, 393.

F. DUCASTELLE, *Order and Phase Stability in Alloys*, North-Holland, Amsterdam (1991).

F. DUCASTELLE and F. CYROT-LACKMANN, *J. Phys. Chem. Solids*, (1970) **31**, 1295; (1971) **32**, 285.

J. FRIEDEL, in *The Physics of Metals, I—Electrons* (ed. J.M. Ziman), Cambridge University Press (1969).

V. HEINE and J. HAFNER, in *Many Atom Interactions in Solids* (M. Puska, ed.), Springer-Verlag, Berlin (1989–90).

D.G. PETTIFOR, in *Metallurgical Thermochemistry*, (O. Kubaschewski, ed.), H.M.S.O. London (1972).

D.G. PETTIFOR, *Phys. Rev. Lett.*, (1989) **63**, 2480.

J.C. SLATER and G.F. KOSTER, *Phys. Rev.*, (1954) **94**, 1498.

A.P. SUTTON, *Electronic Structure of Materials*, Clarendon Press, Oxford (1993).

4

THE COHESIVE ENERGY

4.1 GENERAL METHOD

TO COMPARE carbides of various carbon contents, MC_x, and in various crystal structures a manageable analytical expression is needed for the cohesive energy U in terms of x, of the intermetallic spacing R and of the crystal structure. The one used here models the cohesion by a simple expression which gives most of the cohesive energy by a term U_{unc}; and then, to improve this estimate, adds to it small correction terms so giving U_{corr}.

Although the expression for U_{unc} will be very simple, its inaccuracies will be restricted in two ways. First, its underlying assumptions will be chosen to be consistent with general features of bonding in carbides as set out in the preceding chapters. Second, the values of its coefficients will be chosen to fit measured properties at certain compositions; in particular at $x = 0$ and $x = 1$, representing the pure metal and the stoichiometric carbide respectively. Thus only the (small) changes in U_{unc}, brought about by excursions to other compositions and structures, will be subject to error; and this error can be reduced by the addition of the correction terms.

When comparing the relative stabilities of various compositions of a carbide MC_x, from the values of U_{unc} or U_{corr}, it will be convenient to consider the energy of a standard amount of material expressed in various forms: for example one atom each of M and C in the forms $MC_x + (1 - x)C$; or n atoms of M and one of C in the forms $M_yC + (n - y)M$ with $x^{-1} = y \leqslant n$. The U values of M and C will thus also be needed.

4.2 THE UNCORRECTED ENERGY EXPRESSION

The contribution to U_{unc} from each type of nearest-neighbour bonding will be expressed in the form

$$U = U_{rep} + U_{bond} \qquad (4.1)$$

per atom in the pure metal, or per MC_x unit in the carbide. This is based on the standard Born–Mayer expression for the short-range repulsion between atoms at the distance R,

$$U_{rep}(R) = Ae^{-pR} \qquad (4.2)$$

and on the frequently used (e.g. Friedel, 1969, Ducastelle, 1991) exponential representation of the bonding term,

$$U_{bond}(R) = -Be^{-qR} \qquad (4.3)$$

which is consistent with the exponential decay of atomic wave functions at large distances from their parent nuclei.

In the pure metal, fitting the parameters at $x = 0$ where there is only one type of bonding, i.e. the dd interactions between neighbouring metal atoms, there is one repulsive and one bonding term considered; and thus four parameters, A, p, B, q, to be found. Three relations for these will be obtained by fitting to the observed cohesive energy; to the interatomic spacing from the equilibrium of forces

$$\frac{dU}{dR} = 0 \qquad (4.4)$$

at the equilibrium R; and to the bulk modulus K, from the standard expression

$$K = \frac{R_0^2}{9V_0}\left(\frac{d^2U}{dR^2}\right)_0 \qquad (4.5)$$

where V_0 is the equilibrium atomic volume.

Because the NaCl-type carbides have an FCC metal sublattice it will be convenient to make the above estimates on the assumption that the structure of the pure metal is FCC (with an adjustment, where necessary, in the value of R). The error introduced by this, even when the metal lattice is BCC, is small.

In addition to the above, one other relation is needed to determine the four parameters. Ducastelle (1991) has shown that the best general agreement with experimental observations, for transition metals, is obtained when the value of p/q falls in the range 3 to 5. Accordingly, the value

$$p/q = 4 \qquad (4.6)$$

will be assumed for the pure metals as the required fourth relation.

The corresponding energy expression for the carbides is complicated by the appearance of additional terms; by the reduction relative to the pure metal in the number of electrons available for dd bonding, due to the transfer of some or all of the d electrons to carbon-metal bonds; and by the need to consider different crystal structures, in some cases with a change in coordination.

Because the energy level of the carbon 2s band is so far below that of the valence d electrons, it will be assumed, as a reasonable approximation, that the d contribution to the carbon-metal bonds is limited entirely to pd interactions. It will also be assumed that the direct interaction between neighbouring carbon atoms can be neglected, since its contribution is expected to be too small to justify inclusion in the uncorrected expression. This is suggested by the fact that the characteristic interatomic distance at which carbon atoms interact, e.g. 0.155 nm in diamond and 0.142 in the basal plane of graphite, is far smaller than that between carbon atoms in carbides MC_x ($x \leqslant 1$), e.g. 0.306 in TiC.

There are thus two types of interactions, MM and CM, to be considered in the carbides. A metal atom in, for example, stoichiometric TiC is surrounded by 12 Ti neighbours and 6 C ones, all providing LCAO bonding (and antibonding) orbitals for its electrons. If these orbitals happened all to have the same bond strength h, then we could simply apply the $z^{\frac{1}{2}}$ principle, in the form $(12 + 6)^{\frac{1}{2}}h$, to estimate its bonding. However, the pd orbitals are at a much lower energy level than the dd ones, e.g. by about 6 eV in TiC according to Fig. 2.1, and so are more powerful competitors for the limited supply of valence electrons. Such an effect is implicit in Pettifor's expression, eqn. 3.4, which was used by the writer in a treatment of Fe_3C (Cottrell, 1993).

In strongly bonded carbides such as TiC it is evident from the band structure, e.g. Fig. 2.1, that the pd bonding of the CM interactions completely wins the competition for electrons with the dd bonds of the MM ones, since the pd band is full with 8 electrons per TiC unit, whereas the bonding dd states are completely empty (apart from a small and ambiguous tail near the Fermi level). This allows us to simplify the picture for these carbides by assuming that the carbon atoms 'saturate' their bonds with electrons, i.e. that the pd bonding orbitals of the p band, e.g. in Fig. 2.1, are full; and that in other carbides, such as TiC_x with $x < 1$, those valence d electrons remaining, after saturating the carbon bonds with a contribution of 4 electrons to each carbon atom, go into dd bonds between the metal atoms.

On this basis there are then two distinct sets of U_{rep} and U_{bond} terms in the expression for U_{unc} of a carbide, one for the CM interactions, the other for the MM ones. For the latter it will be assumed that their contribution is given by that of the pure metal adjusted to the R value of the carbide, but with U_{bond} scaled down in proportion to the reduced number of d electrons available for the MM bonds, compared with the pure metal.

On this basis, the cohesive parameters for the pure metal can be inserted into the U_{rep} and U_{bond} terms for the MM contributions to U_{unc} in the carbide. Thus, the uncorrected cohesive energy of NaCl-type MC_x, where the M atom has n valence electrons, can be written as

$$U_{unc} = A'xe^{-p'R} - B'xe^{-q'R} + Ae^{-pR} - B\left(1 - \frac{4x}{n}\right)e^{-qR}, \qquad (4.7)$$

where A', p', B', q' are parameters for the MC bonds, to be determined by experimental fitting, and A, p, B, q are those for the pure metal determined as above. Again, the observed cohesive energy, lattice constant, and bulk modulus of the chosen reference carbide, usually NaCl-type MC, are used for the determination of the parameters. Because the carbon atom is expected to be 'softer' than a transition metal atom, at the spacings in carbides, a somewhat lower value

$$p'/q' = 3 \qquad (4.8)$$

will be assumed. In eqn. 4.7 it is convenient to use R, the metal–metal interatomic distance as the measure of spacing, even though this is not the carbon–metal distance, r_{cm}. For example in NaCl-type carbides, $r_{cm} = R/\sqrt{2}$ so that corresponding to $q'R$ in eqn. 4.7 there is a $q''\,r_{cm}$ where $q'' = q'\sqrt{2}$. In other crystal structures, for example based on the trigonal prismatic cell instead of the octahedral one, a corresponding adjustment has to be made.

No explicit $z^{\frac{1}{2}}$ factor appears in eqn. 4.7. This is partly because the experimental fitting causes it to be absorbed within the numerical values of B and B', although it would appear explicitly when eqn. 4.7 is converted to describe another crystal structure with different coordination numbers. But it also does not appear in the B, B' terms for carbides of other carbon contents, despite the dependence of the total coordination number of the M atoms, $z_{mm} + z_{mc}$, on x. The reason in this case is the division of the total bonding into saturated CM bonds and residual MM ones, as assumed above.

In effect the $\left(1 - \dfrac{4x}{n}\right)$ factor in the final term represents the same

weakening of an MM bond, through the presence of the MC ones, as would a corresponding $z^{\frac{1}{2}}$ factor, although it does so more drastically, in consequence of the assumed over-riding requirement to saturate the CM bonds.

4.3 CORRECTIONS

Whilst U_{unc}, calculated as above, represents all or most of the cohesive energy of a carbide, it is too simply constructed to be able to discriminate effectively between alternative structures and compositions, between their relative stabilities. This discrimination is thus mainly provided by the small secondary effects, with energies generally below 1 eV per atom, which will be represented by the corrections to U_{unc}. There are several of these, although in general not all are important in a particular carbide. Those to be considered are as follows.

4.3.1 *Elastic distortion*

According to the model developed above there are two interpenetrating networks of nearest-neighbour bonds in a carbide; those of the MC and MM bonds, respectively. The spacings at which each of these could be under internal equilibrium, from the forces associated with its own U_{rep} and U_{bond} terms, in general differ from each other and from the overall equilibrium spacing of the carbide. There are thus strains between them. The effect of the strain energy on the cohesion is of course automatically included in the experimentally determined parameters of the particular carbide (usually NaCl-type MC) fitted to experiment. But in, for example, MC_x the strains are expected to produce local distortions round vacant and occupied C-sites. In some cases they make a small but significant contribution to the cohesive properties, which will be estimated.

4.3.2 *Band filling*

It was noted in Chapter 2 that the band structures of the NaCl-type stoichiometric carbides of the group IVA and VA metals are very similar; and can all be represented reasonably well by that of TiC (Fig. 2.1). It follows from Fig. 2.2 that in such carbides of the group VA metals there is a disadvantage due to the raising of the Fermi level to about 2 eV above the minimum in the density of states (where the Fermi level of the group IVA carbides is located). The effect of this is of course built into the experimentally fitted parameters of these carbides.

However, in the non-stoichiometric forms of these carbides additional peaks appear in the density of states distribution, as shown in Fig. 2.4, including a dd bonding peak at only about 1 eV above the minimum. It follows that the change $MC \to MC_x + (1 - x)C$ will benefit in such cases by a fall in the Fermi level which can amount to a significant fraction of an electron-volt, per M atom.

4.3.3 Carbon–carbon repulsion

It was noted in section 1.3 that, where they have the opportunity, carbon atoms prefer sites which maximize the distance between neighbours. This appears to be a general feature in carbides (although sometimes opposed by other effects, discussed later) and suggests that there is a repulsion between nearby carbon atoms. A parallel can be drawn here with the bonding in graphite. The C–C distance between neighbouring hexagonal layers of this, at 0.335 nm, is similar to the C–C spacing in carbides, e.g. 0.306 in TiC; and in both cases the carbon bonding orbitals are all employed elsewhere, i.e. in the $sp^2(\sigma)$ and $p(\pi)$ bonds in the hexagonal layers of graphite; and in the pd bonds with metal atom neighbours in the carbide. Crowell (1958) has given an expression for the energy of interaction of carbon atoms between layers in graphite, which includes the repulsive energy

$$U_{rep} = 1856 \, e^{-35.75 \, R} \qquad (4.9)$$

electron-volts per atom–atom interaction. For example, when $R = 0.3$ nm $U_{rep} = 0.04$ eV. This expression will be used to represent C–C repulsion in the carbides.

4.3.4 Electrostatic energy

The band theory calculations show that in carbides such as TiC and VC there is a charge transfer of about 0.5 e from the metal to the carbon atom (e.g. Neckel et al., 1976). If an NaCl-type carbide were regarded as an 'ionic' crystal from this point of view, the Madelung electrostatic energy due to this charge transfer would be about 1.3 eV per ion. Any such energy is of course already included in the experimental cohesive values fitted to the U_{unc} parameters. However, where opportunities exist, as for example when the crystal structure is changed or when some C-sites are vacant, the system is expected to adjust the local distributions of electrons round the metal, carbon, and vacant, sites so as to gain electrostatic energy. Estimates of significant effects of this kind will be made.

4.3.5 Entropic effects

In non-stoichiometric carbides at high temperatures the carbon atoms are usually in disordered distributions among the C-sites. The free energy due to the ensuing configurational entropy, though small, can be significant. For example, the entropy

$$S = -k\left[\ln x + \left(\frac{1-x}{x}\right) \ln (1-x) \right] \qquad (4.10)$$

provides a free energy gain of about 0.22 eV per C atom in $MC_{0.5}$ at 2000 K. This and other small entropic contributions will be included where necessary.

REFERENCES

A.H. COTTRELL, *Mater. Sci. Tech.*, (1993) **9**, 227.

A.D. CROWELL, *J. Chem. Phys.*, (1958) **29**, 446.

F. DUCASTELLE, *Order and Phase Stability in Alloys*, North-Holland, Amsterdam (1991).

J. FRIEDEL, in *The Physics of Metals, I—Electrons*, (J.M. Ziman, ed.), Cambridge University Press (1969).

A. NECKEL, P. RASTL, R. EIBLER, P. WEINBERGER and K. SCHWARZ, *J. Phys. C., Sol. State Phys.*, (1976) **9**, 579.

TITANIUM CARBIDE

5.1 THE UNCORRECTED COHESIVE ENERGY

THE FIRST step in applying the method of Chapter 4 to titanium carbide is to determine the parameters in eqn. 4.7 for the uncorrected cohesive energy. This is done in two stages. First for the pure metal, assumed to be FCC with $x = 0$. Second, for the stoichiometric TiC, with $x = 1$ and $n = 4$. The experimental values used to fit these parameters are the cohesive energy, the equilibrium intermetallic distance R and the bulk modulus K.

The values of the cohesive energy U, used here, are measured relative to the free neutral atoms in their ground states. In general the electronic structure of an atom has to be promoted to a suitable excited state, e.g. $4s^2 3d^2 \rightarrow 4s^1 3d^3$ for titanium, before it forms bonds in a molecular or condensed state. In some calculations U is measured relative to this promoted atomic state (Gelatt et al., 1977; Cottrell, 1994). However, it may reasonably be assumed here that the promotion energies remain unchanged through the various metal–carbon combinations to be considered and thus have no effect on their relative stabilities. Accordingly, U will be measured here from the unpromoted atomic (ground) energy levels. In fact, the values of R_{unc} and relative stabilities obtained in this way for TiC$_x$ (and the carbides of other metals) differ imperceptibly from those deduced from promoted atomic energy levels (Cottrell, 1994).

Using the experimental values $U = -4.85$ eV, $R = 0.293$ nm and $K = 1.05 \times 10^5$ MNm^{-2} = 656 eVnm^{-3} for titanium and $U = -4.85 - 7.38$ (graphite) $- 1.91$ (heat of formation) $= -14.14$, $R = 0.3062$ and $K = 1500$ for TiC, the equation for the uncorrected cohesive energy becomes, per TiC$_x$ unit

$$U_{unc} = 8810 x e^{-24.1R} - 242 x e^{-8.03R}$$
$$+ 17900\, e^{-31.78R} - 66.28(1 - x)\, e^{-7.94R} \tag{5.1}$$

Table 5.1 gives values of R_{unc} and U_{unc} obtained from this equation for various non-stoichiometric carbides.

As x drops below unity, the values of R_{unc} do not reproduce the initial slight increase in the observed R and generally fall lower than the latter. This will be discussed further when the corrections are made. The values of U_{unc} in the table are for 6 Ti + 1 C atoms, expressed in the various forms Ti_yC + $(6 - y)Ti$, where $xy = 1$. It is striking how little this energy changes until the carbon content is diluted by increasing y substantially. This is mainly because the local environment of a carbon atom remains constant throughout, at $z_{cm} = 6$, apart from the small effect of the change in R, so that the C-Ti contribution to U_{unc} is, per C atom, virtually unchanged. Furthermore, the assumed form of the Ti–Ti bonding contribution in eqn. 5.1, with its $(1 - x)$ dependence, makes this residual contribution almost independent of y. The form of eqn. 5.1 is thus such that U_{unc}, per 6M + 1C, depends only on the minor variations of bond and repulsive energies with the small changes of R. This near constancy of U_{unc} opens the way for various second-order contributors to exert significant effects, discussed later.

5.2 EQUILIBRIUM OF FORCES

The $(1 - x)$ factor in eqn. 5.1, which represents the fraction of d electrons remaining in M–M bonds, shows that in stoichiometric TiC the Ti–Ti forces are entirely repulsive. Taking $x = 1$ and dividing by six, the terms in this equation give the various individual bond energies in TiC, from which the bond forces can be found. The Ti–Ti force is

$$f_{Ti - Ti} = \frac{d}{dR}\left[2983e^{-31.78R}\right] = -94810\,e^{-31.78R} \tag{5.2}$$

and the corresponding attractive force in a C–Ti bond is

$$f_{C - Ti} = \frac{dR}{dr}\frac{d}{dR}\left[1468\,e^{-24.1R} - 40.3\,e^{-8.03R}\right]$$

$$= -50033\,e^{-24.1R} + 457.7\,e^{-8.03R} \tag{5.3}$$

where $r = R/\sqrt{2}$ = the C–Ti distance. Thus, in equilibrium

$$f_{C-Ti} + \sqrt{2}\,f_{Ti-Ti} = 0 \tag{5.4}$$

The condition $f_{C-Ti} = 0$ applied to eqn. 5.3 defines the spacing, $R = 0.292$, at which the C–Ti bond network would be in internal equilibrium. The corresponding $r = 0.206$ is near to the sum of the standard covalent radii of carbon (0.077) and titanium (0.132). It is striking that carbon expands the lattice of titanium, despite the strong affinity between them and

Table 5.1 Cohesive parameters of titanium carbides. The values of U_{unc} are per 6 Ti + 1 C atoms. Bracketed values are experimental fits.

	TiC	Ti$_{1.1}$C	Ti$_{1.2}$C	Ti$_{1.5}$C	Ti$_2$C	Ti$_3$C	Ti$_6$C
x	1	0.917	0.833	0.667	0.5	0.333	0.167
R_{obs}	0.3062	0.3062	0.3063	0.3055	0.3039	–	–
R_{unc}	(0.3062)	0.3058	0.3054	0.3044	0.3030	0.3010	0.2980
U_{unc}	(38.39)	38.39	38.37	38.36	38.32	38.29	38.24

the tendency of the C–Ti bond to contract. This is an indication of the weakening of the Ti–Ti bond forces, virtually to zero, by the transfer of the d electrons into the C–Ti bonds, as was surmised by Rundle (1948) from a comparison of interatomic distances in NaCl-type MC carbides (M = Ti, Zr, Hf etc.).

As x is reduced the Ti–Ti bonds acquire electrons for dd bonding, which opposes the repulsive force in eqn. 5.2. Thus,

$$f_{\text{Ti–Ti}} = -94810\,e^{-31.78\,R} + 87.71(1 - x)\,e^{-7.94\,R} \tag{5.5}$$

which, at the lowest observed limit of this phase, i.e. $x = 0.5$ and $R = 0.3039$, is negative. Thus, over the whole range of the phase neighbouring titanium atoms repel and the C–Ti bonds are in tension.

5.3 ELASTIC DISTORTION

An obvious correction is suggested by these results. In non-stoichiometric TiC$_x$ the empty C-cells expand, since their constraining C–Ti tensions have been removed, and this expansion is resisted elastically by the neighbouring material. The expansion has been treated as spherically symmetrical (Cottrell, 1994) and subsequently as non-symmetric (Cottrell, 1995). The second of these will now be followed.

Consider TiC$_{0.5}$ with a random distribution of filled ('C-cells')and empty ('E-cells') carbon cells. The distortion will be described in terms of the deviation of Ti–Ti nearest-neighbour distances from their R_{unc} value, 0.303. Each Ti–Ti bond passes between two such cells and so its length can be written as R_{CC}, R_{CE}, R_{EE}, according to the types of these. It will be assumed that the distorted structure can be described in terms of three values for R_{CC}, R_{CE} and R_{EE} which are fixed for a given x.

The average Ti atom in TiC$_{0.5}$ is joined, along the cube axes of the crystal, to 3 carbon neighbours. There are two types of such a configuration. In the 'A-type' the joins at the ends of one cube axis are both to C atoms, those for another axis are both E and the third has one C, one E. In the 'B-type' each axis has one C, one E. Statistically, the A-type is expected to be 1.5 times as abundant as the B-type. A Ti atom in the A-type has two R_{CC} links, eight R_{CE} and two R_{EE}, i.e. it has a 2/8/2 set of links. In the B-type it has a 3/6/3 set.

Obviously, $R_{CC} < R_{CE} < R_{EE}$ since the Ti atoms of a CC link can, by reducing R_{CC}, get closer to the adjoining C atoms, i.e. can reduce R_{CTi} $(=\sqrt{2}r)$ and so improve the cohesion. Since these various R values develop from the common starting point, R_{unc}, it is assumed for simplicity that $R_{CE} = R_{unc}$. It is also assumed that

$$R_{CTi} = \tfrac{1}{2}(R_{CC} + R_{CE}) \tag{5.6}$$

and that

$$\text{A-type: } R_{corr} = \tfrac{1}{12}(2R_{CC} + 8R_{CE} + 2R_{EE}) \tag{5.7}$$

$$\text{B-type: } R_{corr} = \tfrac{1}{12}(3R_{CC} + 6R_{CE} + 3R_{EE}) \tag{5.8}$$

The method is to assume various pairs of R_{CC} and R_{corr} values and, from these and also $R_{CE} = R_{unc} = 0.303$, to evaluate the total bond energy. The elastic misfit energy is also estimated and the sum minimized to give R_{corr} and U_{corr}.

The elastic energy, which opposes the tendency of R_{EE} to expand and R_{CC} to contract, is due partly to the difference in volumes of the E-cells and C-cells and partly to the distortion of the octahedra by the unequal edge lengths. They can be combined approximately into a single model of 'tetragonal' distortion by the trigonal antiprism representation of the octahedral cell, as shown in Fig. 5.1, similar to that proposed by Zener

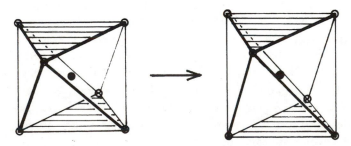

Fig. 5.1 Distortion of an octahedron, regarded as a trigonal antiprism.

(1946) in his elastic theory of tetragonal martensite. The tetragonality strain is assumed to be

$$\epsilon_C = (R_{CE} - R_{CC})/R_{CE} \qquad (5.9)$$

$$\epsilon_E = (R_{EE} - R_{CE})/R_{CE} \qquad (5.10)$$

in the C-cells and E-cells, respectively. The strain energy per two cells (one C, one E), of total volume Ω (= 0.0265) is then taken as

$$U_E = \tfrac{1}{4}(\epsilon_C^2 + \epsilon_E^2)\Omega Y \qquad (5.11)$$

where Y = Young's modulus $\simeq$ 1970, which is 0.7 of the value for TiC (Toth, 1967).

Separate optimizations of the total (bond + elastic) energies for the two configurations then lead to R_{corr} = 0.3035 (A-type) and 0.3039 (B-type) and thus to an average R_{corr} = 0.3037 for TiC$_{0.5}$. This result, together with corresponding ones for other compositions, is compared in Fig. 5.2 with observed values. On this interpretation the initial small lattice expansion, which occurs as x drops below unity, is due to the loss of the tensile forces in those octahedra which become E-cells. The contraction which sets in as x drops further is of course due to the increase of dd bonding forces between titanium atoms.

5.4 CORRECTED COHESIVE ENERGIES

Corresponding to the above corrections in R, values of the correction $-\Delta U_1$ to U_{unc} have been estimated, in Table 5.2. Since the number of carbon–carbon neighbours varies with x, it is necessary to add a correction ΔU_2 for the C–C repulsion given by eqn. 4.9. The value for TiC, 0.196 per carbon atom, is already contained within the fitted (observed) value of the cohesive energy. Those for $x < 1$ are smaller and so their values appear in Table 5.2 as gains in cohesion.

The resulting U_{corr} values in the Table are remarkable for the smallness of their variation over the composition range. Since they refer to carbides in the presence of excess titanium their maximum should locate the position of the Ti-rich boundary of the TiC phase field. This lies in the region Ti$_{1.5}$C to Ti$_2$C, which is consistent with the high-temperature location in the phase diagram, Fig. 5.3. However, U refers to low temperatures and, although the phase diagram is not well established there, it is evident from Fig. 5.3 that

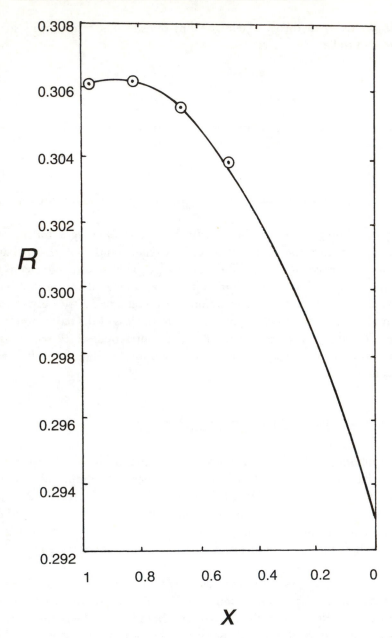

Fig. 5.2 Variation of Ti–Ti interatomic spacing in TiC_x alloys. Circled points are from lattice parameter measurements (Timofeeva and Klochkov, 1974). The curve follows R_{corr} values.

Table 5.2 Corrected cohesive energies (eV) per 6 Ti + 1C atoms.

	TiC	Ti$_{1.1}$C	Ti$_{1.2}$C	Ti$_{1.5}$C	Ti$_2$C	Ti$_3$C	Ti$_6$C
$-U_{unc}$	38.39	38.39	38.37	38.36	38.32	38.29	38.24
$-\Delta U_1$	0	0.01	0.02	0.02	0.03	0.03	0.02
$-\Delta U_2$	0	0.02	0.03	0.06	0.09	0.11	0.15
$-U_{corr}$	38.39	38.42	38.42	38.44	38.44	38.43	38.41
At 1190 K:							
$-TS_1$	0	0.031	0.054	0.095	0.138	0.19	0.269
$-TS_2$	0.048	0.047	0.046	0.043	0.038	0.03	0
$-G$	38.44	38.50	38.52	38.58	38.62	38.65	38.68
At 1920 K:							
$-TS_1$	0	0.05	0.086	0.153	0.222	0.306	0.433
$-TS_2$	0.029	0.029	0.028	0.026	0.023	0.017	0
$-TS_3$	0.115	0.113	0.110	0.103	0.092	0.069	0
$-G$	38.53	38.61	38.64	38.72	38.78	38.82	38.84

the boundary does not extend down to such low carbon contents. The near equality of all the U_{corr} values shows, of course, that even a minor inaccuracy in the estimation could shift the position of the maximum substantially.

To extend the consideration to high temperatures, some estimates of entropic contributions are included in Table 5.2, at the two critical temperatures 1190 K (upper limit of α-Ti) and 1920 K (β-Ti eutectic). The first derives from the entropy of mixing S_1, assuming a random distribution of the carbon atoms in their octahedral sites and using eqn. 4.10. The second represents the gain in free energy of the excess, free, titanium, due to the entropy of mixing of the primary solution of C in Ti, assuming for this values of x from Fig. 5.3. The third, $-TS_3$, coming from the extra entropy of β-Ti, is about 0.035 eV per Ti atom, derived from the latent heat of the α–β transition. As expected, these entropic contributions tilt the equilibrium towards low-carbon compositions. The difference between the estimated free energies of Ti$_2$C, which is approximately where the observed phase boundary occurs, and those of the hypothetical lower carbides, Ti$_3$C and Ti$_6$C, remains very small.

The position of the high-carbon phase boundary can be estimated by considering the relative stabilities of carbon-rich mixtures, as in the 1 Ti + 1 C series, in the forms TiC$_x$ + (1 − x)C, given in Table 5.3. Both U_{corr} and its entropically adjusted value, at 1920 K, indicate that the stoichiometric composition should be the most stable. In practice, as Fig. 5.3 shows, the

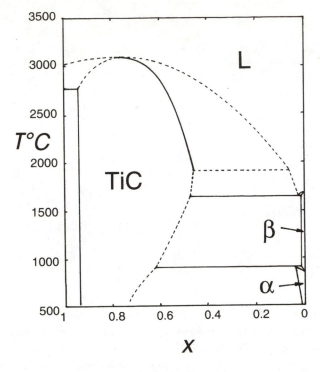

Fig. 5.3 The titanium–carbon phase diagram. The composition variable is represented by x in TiC$_x$.

limit appears to be at $x \simeq 0.95$. It has been suggested (Storms, 1967) that this may be due to the difficulty of eliminating all oxygen, which is a rival candidate for the octahedral sites.

5.5 PRIMARY SOLUBILITY

The phase diagram shows that the TiC phase comes into equilibrium with the primary solution of carbon in titanium; but both the α-phase (CPH), below 1190 K, and the β-phase (BCC), from 1190 to 1943 K, have small solubility.

   Consider a single carbon atom in an interstitial site in (FCC) titanium. It will be assumed that the site retains its octahedral symmetry, although of different size from those of the surrounding metal. The six metal atoms forming this octahedron make bonds of three different lengths: $R_{CTi}/\sqrt{2}$ with

Table 5.3 Corrected cohesive energies (eV) per 1 Ti + 1C atoms.

	TiC	TiC$_{0.91}$	TiC$_{0.83}$	TiC$_{0.67}$	TiC$_{0.5}$	TiC$_{0.33}$	TiC$_{0.17}$
$-U_{corr}$	14.14	14.00	13.85	13.54	13.21	12.88	12.55
$-TS_1$	(1920 K)	0.05	0.07	0.10	0.11	0.10	0.07
$-G$	14.14	14.05	13.92	13.64	13.33	12.98	12.62

the carbon atom; R_{TiTiC} with neighbouring atoms of the octahedron; and R_{TiTiE} with those outside it. Each such atom contributes $\frac{2}{3}$ of an electron to the CTi bonds, so that its $(1 - x)$ in eqn. 5.1 is 0.8333. Hence the energy of the octahedron is

$$U = 8810\, e^{-24.1\, R} - 242\, e^{-8.03\, R} + 2(17900\, e^{-31.78\, R} - 55.23\, e^{-7.94\, R}) \quad (5.12)$$

which gives $-U = 23.03$ at the internal equilibrium R = 0.2958. Since in their external TiTi links the octahedral metal atoms share their bonding with atoms not depleted in electrons, it will be assumed that $(1 - x)$ for these is $\frac{1}{2}(1 + 0.8333) = 0.917$. There are 48 such links so that their energy is

$$U = 8(17900\, e^{-31.78\, R} - 60.78\, e^{-7.94\, R}) \quad (5.13)$$

giving $-U = 34.6$ at $R = 0.2967$. This bond energy is shared with the external atoms so that only half of it 'belongs' to the internal ones, thus giving a total $-U = 40.33$ for the octahedron. The weakening of the external bonds has also to be included. At full strength, at the equilibrium $R(0.293)$ of the pure metal, their half share of $-U$ would have been 19.42. Hence their 17.3 implies an energy penalty of 2.12 due to the presence of the carbon atom, so that the final value for this atom and its 6 metal neighbours, before including elastic energy, is 38.21. This is 0.11 less cohesive than Ti$_2$C + 4 Ti, according to the U_{unc} value in Table 5.2; and 0.41 less, according to the G value.

The above R values represent an expansive strain, $\epsilon \simeq 0.005$, relative to pure titanium, of a region extending out to about $1.4\,R$ from the carbon centre. Assuming a shear modulus $\mu \simeq 400$ the standard elementary theory (e.g. Cottrell, 1988) gives an elastic energy

$$8\pi\mu(1.4\,R)^3\epsilon^2 \simeq 0.02 \quad (5.14)$$

so that the total energy penalty for a carbon atom to leave Ti$_2$C and enter dilute solution in (FCC) titanium is $\Delta H \simeq 0.43$. The solubility limit

$$c \simeq e^{-\Delta H/kT} \qquad (5.15)$$

where c is atomic concentration, is thus of order 0.01 at 1190 K, which is comparable with the observed value for α-Ti (Fig. 5.3). Of course the latter is CPH, not FCC, but its octahedral sites are very similar. The lower solubility in β-Ti is presumably due to the less suitable interstitial sites in the BCC crystal structure.

5.6 CRYSTAL STRUCTURE

In the uncorrected approximation the TiC structures with FCC and (ideal) CPH metal sublattices have equal cohesive energies, since they depend only on the nearest-neighbour pairwise interactions which are the same in both, at the same R. The preference for FCC over CPH thus stems from the correction effects.

There is a difference in the sublattices of octahedral carbon sites; that of the cubic carbide being FCC, so that each carbon atom has twelve carbon neighbours, all at a distance $R = 0.3062$, whereas in the hexagonal structure the carbon sublattice is simple hexagonal with six neighbours in the basal plane at $R = 0.3062$ and two, vertically above and below, at $R\sqrt{2/3} = 0.250$. The closeness of the latter gives the hexagonal structure a larger C–C repulsion, 0.342 eV per carbon atom according to eqn. 4.9, whereas in cubic TiC it is 0.196, so that the hexagonal structure is disadvantaged by 0.146 eV per TiC unit through this effect.

In non-stoichiometric TiC_x, in which the carbon atoms are randomly distributed among the octahedral cells, the same disadvantage for the hexagonal structure continues, at a correspondingly reduced level. However, when $x \ll 1$ there is opportunity for the carbon atoms to avoid being close neighbours by taking up ordered distributions, with of course a loss of entropy. These are discussed below.

In $TiC_x (1 > x > 0.5)$ and the corresponding carbides of the other group IV A metals, the NaCl-type structure prevails over all others. The BCC alternative, a TiC 'martensite', is presumably ruled out because of its less efficient packing, due to the necessity to push apart two of the six M atoms of the cage round the central C, in order to make an octahedral site.

A WC-type structure for TiC, or indeed any structure which encloses the carbon atoms in trigonal prismatic cages instead of octahedral ones, would require a smaller Ti–Ti spacing, R, for a given C–Ti one. Per unit C–Ti spacing, $R = 1.309$ for the trigonal prism and 1.414 for the octahedron.

Since the forces between neighbouring titanium atoms in TiC are repulsive, this disadvantages the WC-type structure.

As pointed out in Chapter 1, complex structures such as $M_{23}C_6$ have a lower density of sites for carbon, compared with the above, and so are disfavoured compared with the latter when strong carbide-forming metals such as titanium are equilibrated with excess carbon. The Cr_3C_2-type and Cr_7C_3-type structures, as alternatives to cubic $TiC_{0.67}$ or $TiC_{0.43}$ in the presence of excess titanium, are presumably ruled out because their trigonal prismatic metal cells would, as in the case of the WC-type alternative, aggravate the Ti–Ti repulsion.

5.7 ELASTIC CONSTANTS

Because they involve a double differentiation of energy with respect to strain, the elastic constants provide a sensitive test of the fine detail of a theory of cohesion. In the present case, of course, the bulk modulus, $K = \frac{1}{3}(c_{11} + 2c_{12})$, is excluded from this since it is experimentally fitted.

The standard expressions for the elastic constants (e.g. Mott and Jones, 1936), applied to U_{unc} (eqn. 5.1), give the values in Table 5.4. Since U_{unc} is based on central forces the Cauchy pressure, $c_{12} - c_{44}$, is necessarily zero. Comparison with the measured values (Gilman and Roberts, 1961) shows that the U_{unc} expression substantially underestimates c_{44}. The small value of this stems from the fact that shear on (100) planes leaves the strong C–Ti bonds largely unstrained. By contrast the other shear constant $(c_{11} - c_{12})/2$, in which these bonds are significantly involved, is predicted to be quite large by the theory. The discrepancy cannot be attributed to the correction factors of Chapter 4, since these do not apply to cubic stoichiometric titanium carbide. They indicate an incompleteness in the form of eqn. 5.1 as a model of the cohesion, even though this equation has been fitted to give the observed values of cohesive energy, lattice constant and bulk modulus. A pointer to the missing feature is provided by the observed failure of the Cauchy relation, i.e. $c_{12} - c_{44} = -388$, which indicates the presence of a

Table 5.4 Elastic Constants of Titanium Carbide, $eVnm^{-3}$

	C_{11}	C_{12}	C_{44}	$(C_{11}-C_{12})/2$
Observed	3125	706	1094	1209
From U_{unc}	3602	457	457	1572

non-central force; and, from its negative value, a force characteristic of brittle crystals, i.e. those with $\mu/K > 0.5$.

This force is almost certainly due to the angular character of the orbitals involved in the pd bonding. A key feature here, in the words of Pettifor (1992), is that 'it is the σ, π or δ bond which is being embedded in the local environment, rather than a spherically symmetric atom'. Such an angular effect would be produced by the four-hop paths that contribute to μ_4 in the moments theory, e.g. those starting from atoms d, e, f in Fig. 3.2. Its contribution to the cohesive energy is expected to be small (see section 5.8 below). In fact, an angular effect which added the needed $600\,eV\,nm^{-3}$ to the calculated c_{44}, in Table 5.4, would change the cohesive energy by only $0.06\,eV$ even at the large shear strain of 0.1.

5.8 ORDERED STRUCTURES

Neutron spectroscopy has established that when $x < 0.67$ and $T < 1000\,K$ the disordered distribution of carbon atoms among the octahedral sites, in TiC_x, is replaced by an ordered one (Goretski, 1967; Moisy-Maurice et al., 1982). It is simplest to consider $TiC_{0.5}$, in which case there is a 50/50 distribution of carbon atoms and vacancies on the FCC sublattice. Two forms of order on this sublattice have been observed, as shown in Fig. 5.4; the cubic Fd3m type in which the (1 1 1) planes are alternately one-quarter

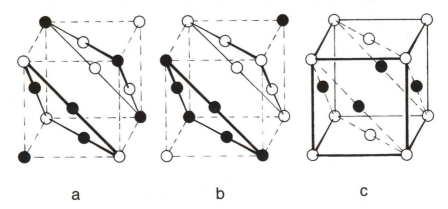

a   b   c

Fig. 5.4 Positions of carbon atoms (filled circles) and vacant sites (empty circles) in the FCC sublattice of ordered $TiC_{0.5}$: (a) the Fd3m structure; (b) the R3̄m (CuPt-type) structure. Also shown (c) is a hypothetical structure of CuAu-type.

and three-quarters filled with carbon atoms; and the $R\bar{3}m$ type, typified by the CuPt superlattice, a layer structure with alternately full and empty (1 1 1) planes, slightly distorted along the < 1 1 1 > axis to a rhombohedral form. In both structures each titanium atom has three carbon neighbours (along < 1 0 0 > axes) on one side and three vacancies on the other side. The observations suggest that the Fd3m structure forms first but gradually converts to the CuPt-type during long annealing. The ordering temperature is a maximum, 1058 K, in $TiC_{0.63}$, but is fairly insensitive to composition in the range $0.67 > x > 0.58$ (Moisy-Maurice et al., 1982). Also shown in Fig. 5.4 is the CuAu-type structure, which might have been expected in $TiC_{0.5}$, but in fact is not formed.

A detailed theory of such ordering has been given by Landesmann et al. (1985), based on band energy estimates, as in eqn. 2.4. They showed that the energy of $MC_{0.5}$ is lowest for the CuPt-type of ordering (although hardly different from that for the Fd3m type), but the calculations were difficult because the energy advantage of this ordered structure over the disordered one was only about one-hundredth of the total band energy. Nevertheless, this was an overestimate because the observed ordering temperature indicated an even smaller advantage. The key feature brought out by these calculations was an indirect repulsive interaction between second neighbour carbon atoms, along < 1 0 0 > axes. The interaction was indirect because it involved electronic four-hops (i.e. μ_4 in section 3.4) via the intermediate metal atom, there and back along the cube axis, in a straight line, as in Fig. 3.2(f).

We shall here not follow this method but instead seek appropriate corrections to the basic U_{unc} expression. The linear dependence on x of the terms in this, eqn. 5.1, implies that in the absence of corrections and at fixed R the energy is independent of the distribution of the carbon atoms. It also means that the uncorrected expression can be interpreted as a homogeneous model of, for example, $TiC_{0.5}$, in which every octahedral site is occupied by a 'half carbon atom'; half in the sense that it forms only half as many bonds with its six titanium neighbours. The remaining d electrons, surplus to this bonding, are in this model distributed equally among all the Ti–Ti links, each atom contributing $\frac{1}{6}$ of an electron to each of its twelve such links.

The corrections then result from the pairwise condensation of these 'half atoms' into full carbon atoms in half of the octahedral sites, so producing a disordered or ordered, atomically heterogeneous, structure. When a half-atom is moved into the site of a neighbouring one, it takes with it two electrons from the titanium atoms, to form its C–Ti bonds. If the electrons in the Ti–Ti bonds remained equally distributed, as before, there would then result an imbalance of electronic charge, with excess in the filled octahedra and a deficit in the vacated ones. A partial redistribution of the dd electrons is thus expected, away from CC-type Ti–Ti bonds (cf section 5.3) and into the EE ones, with the electron content of the CE links remaining fairly unchanged.

Conditions are most favourable for this redistribution when each titanium atom has 3 CC and 3 EE links, on opposite sides, separated necessarily by a sheet of 6 CE links, i.e. a 3/6/3 arrangement. This occurs in the ordered structures but not much in a random one, where the numbers of carbon neighbours to a titanium atom often differ from three. According to Bernouilli statistics about 0.31 of the titanium atoms have 3 carbons; 0.48 are equally likely to have 2 or 4; and 0.18 similarly 1 or 5. When for example there are four neighbours a titanium atom can have at most only one EE, i.e. only one-third of the number of electron 'sinks' available to a 3/6/3 atom. Similarly for the number of 'sources' when there are only two carbons. On the assumption that the electron redistribution about a titanium atom is proportional to the number of its minority links, sources or sinks, the average over the Bernouilli population indicates that the electronic redistribution in the random case should average about half of that in the ordered one.

The CuPt-type superlattice in ordered $TiC_{0.5}$ is a simple FCC stacking of $(1\,1\,1)$ close-packed sheets in the sequence TiCTiETiCTiE ... Each titanium atom thus has 3 CC links into its adjoining C layer, 3 EE into its adjoining E layer and 6 CE within its own $(1\,1\,1)$ layer. According to band theory calculations (Schwarz and Neckel, 1986; Redinger et al., 1985) there is a charge transfer of about 0.5 e to a carbon atom in TiC. It is thus assumed that in the 3/6/3 case 0.25 e is transferred from the three CC to the three EE, i.e. that the effective value of $(1 - x)$ in eqn. 5.1 is changed from its value of 0.5 in the homogeneous model of $TiC_{0.5}$ to 0.25 for CC links, 0.5 for CE ones and 0.75 for EE. Assuming also that $R_{CE} = \frac{1}{2}(R_{CC} + R_{EE})$ the equation can then be used to calculate the equilibrium values, $R_{CC} = 0.302$, $R_{CE} = 0.304$, $R_{EE} = 0.307$, and $-U = 9.48$.

The corresponding value for the disordered structure, from eqn. 5.1, is $-U = 9.46$, which, after the corrections for different R_{CC}, R_{CE}, R_{EE} values as in section 5.3, improves to $-U = 9.48$. Since the average carbon atom has three carbon nearest neighbours in both the disordered and ordered states and since the average CC spacing is about the same, the CC-repulsion energy is also expected to be about the same in both.

Hence, the sum of all the above contributors produces no ordering energy, which leads to the conclusion that, in this system, the electrostatic correction energy is, to the present approximation, entirely responsible for ordering. The shift of about 0.5 e from the CC Ti–Ti links, at a distance of about $r \simeq 0.5\,R$ from the carbon atoms, to the EE links where $r \simeq R$, releases electrostatic energy assumed to be given approximately by the screened Coulomb expression,

$$U_E = \frac{q^2}{r}\,\mathrm{e}^{-r/r_0} \qquad (5.16)$$

where r_0 is the screening length and q is the interacting charge (0.5 e). The value of r_0 is considered to be about 0.05 nm for electronically dense metals (Herring, 1966; Smith, 1975). Using this and the two values of r, in eqn. 5.16, the gain from the electronic redistribution in the ordered state is 0.11 eV. Since the disordered gain is expected to be about half of this, the ordering energy is estimated to be about $Q \simeq 0.055$ eV per titanium atom. The corresponding order–disorder transition temperature T_c is then given approximately by

$$k\,T_c\,\ln 2 \simeq Q \qquad (5.17)$$

i.e. $T_c \simeq 960$ K which compares reasonably with the observed value.

The corresponding calculation for the Fd3m structure gives essentially the same results. The main difference is that the cubic symmetry of this precludes the independent optimisation of R_{CC} and R_{EE}. As a result, this structure suffers a small strain energy disadvantage, estimated to be about 0.002 eV per titanium atom, relative to the CuPt-type (Cottrell, 1995).

The above electrostatic effect provides a simple explanation of why $TiC_{0.5}$ does not order in a CuAu-type structure (Fig. 5.4). In this the carbon and vacant sites would separate on to alternate (100) planes. A titanium atom must then have either four carbon neighbours, i.e. with CC/CE/EE = 4/8/0; or two, with CC/CE/EE = 0/8/4. There is thus little scope for electronic redistribution since there are either no main electron 'sources' (CC) or 'sinks' (EE).

REFERENCES

A.H. COTTRELL, *Introduction to the Modern Theory of Metals*, The Institute of Metals, London (1988).

A.H. COTTRELL, *Mater. Sci. Tech.*, (1994) **10**, 22.

A.H. COTTRELL, *Mater. Sci. Tech.*, (1994) **10**, 788.

A.H. COTTRELL, *Mater. Sci. Tech.*, (1995), in press.

C.D. GELATT, H. EHRENREICH and R.E. WATSON, *Phys. Rev. B*, (1977) **15**, 1613.

J.J. GILMAN and B.W. ROBERTS, *J. Appl. Phys.*, (1961) **32**, 1405.

H. GORETSKI, *Phys. Stat. Solidi*, (1967) **20**, K 141.

C. HERRING, in *Magnetism IV*, (G.T. Rado and H. Suhl eds), Academic Press, New York (1966).

J.P. LANDESMANN, G. TRÉGLIA, P. TURCHI and F. DUCASTELLE, *J. Physique* (1985) **46**, 1001.

V. MOISY-MAURICE, N. LORENZELLI, C. H. DE NOVION and P. CONVERT, *Acta. Met.*, (1982) **30**, 1769.

N.F. MOTT and H. JONES, *The Theory of the Properties of Metals and Alloys*, Oxford University Press (1936).

D.G. PETTIFOR in *Electron Theory in Alloy Design*, (D.G. Pettifor and A.H. Cottrell eds), The Institute of Materials, London (1992).

R.E. RUNDLE, *Acta. Cryst.*, (1948) **1**, 180.

J.R. SMITH, in *Interactions on Metal Surfaces*, (R. Gomer ed.), Topics in Applied Physics, **4**, Springer-Verlag, Berlin (1975).

E.K. STORMS, *The Refractory Carbides*, Academic Press, New York (1967).

I.I. TIMOFEEVA and L.A. KLOCHKOV, in *Refractory Carbides* (G.V. Samsonov ed.), Consultants Bureau, New York and London (1974).

L.E. TOTH, *Transition Metal Carbides and Nitrides*, Academic Press, New York (1971).

C. ZENER, *Trans. Amer. Inst. Min. Met. Eng.*, (1946) **167**, 550.

6

THE EARLY TRANSITION METALS

6.1 GENERAL CHARACTERISTICS

THERE ARE many similarities in carbide formation between titanium and the other early transition metals of the periodic table. The MC_x phase, with $0.5 \leqslant x \leqslant 1$ and the NaCl-type structure, is the dominant carbide in the group IV A and V A metals, although the latter also form a carbide at $x \simeq 0.5$ with a CPH metal sublattice. The group III A metals are less similar. Scandium and yttrium form a non-stoichiometric carbide with a CPH metal sublattice over a wide composition range centred about $x \simeq 0.5$ at low temperatures, although it extends up to and beyond $x = 1$ in YC at 2000 K. These metals also form several carbides with other structures at x values both below and above $x = 1$. Lanthanum combines with excess carbon to form carbides of other structures, MC_x where $x = 1.5$ or 2.

Electronic calculations (Chapter 2) show that all the NaCl-type stoichiometric carbides have practically the same band structure as that of TiC (Fig. 2.1), so that a rigid-band assumption can reasonably be made, enabling Fig. 2.1 to be used for all, with appropriate placing of the Fermi level, E_F. The band structure energy is most favourable for the Ti, Zr, Hf group since here E_F falls in the minimum in the density of states distribution, between the bonding and antibonding states. Fig. 2.2 shows that the density of states at E_F rises sharply when the number of valence electrons per MC unit is reduced from 8 to 7, as in group III A stoichiometric carbides. Because of this, the fall in E_F is insufficient to compensate for the reduction in bonding electron numbers and so these group III A carbides have smaller band structure energies.

By contrast, since the pd bonds are saturated at 8 electrons per MC unit, the extra electron per metal atom of the group V A carbides has to form the somewhat weaker dd bonds between M atoms, as indicated by the density of states peaks at higher energies in Fig. 2.2. Thus the band structure energy is also smaller for these group V A carbides.

These band structure energies do not uniquely determine the heats of carbide formation (Table 2.1) because the latter are measured relative to the energies of formation of the constituent elements in their standard states. Thus the elemental metals have to be 'de-bonded', almost completely for groups III A and IV A and very substantially for group V A, to provide electrons for the C–M bonds. Differences in the cohesive energies of the metals thus contribute to the relative heats of formation in Table 2.1.

6.2 CARBON–CARBON BONDING

When, as with the group III A carbides above, the metal atoms cannot supply enough valence electrons to saturate the pd bonds, it becomes advantageous to reduce the number of p states provided for them by the carbon atoms. This can be done in two ways. First, obviously, by reducing the carbon content as in the carbides, MC_x with $x \simeq 0.5$, of scandium and yttrium.

Second and more interestingly by internalising some of the carbon p states and electrons into carbon–carbon bonds. This requires nearest-neighbour C–C distances to be drastically reduced into the range, about 0.13 nm, at which C–C covalent bonding occurs. As a result, structures are formed in which close carbon pairs, which resemble C_2 molecules, occupy the interstitial sites. Typical are the bicarbides, YC_2, LaC_2 (and also UC_2) which, although metallic, have a structure similar to that of the non-metallic CaC_2. As shown in Fig. 6.1 it can be regarded as an NaCl-type structure in which single C atoms in the octahedral sites have been replaced by C_2 pairs, the axes of these being all parallel to the same cubic axis, so distorting the structure to tetragonal in this direction. The non-metallic compound CaC_2, with a C_2

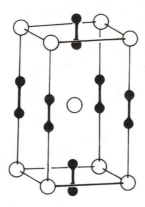

Fig. 6.1 Tetragonally distorted structure of bicarbides.

spacing of 0.12 nm, is of course readily decomposed by water to form acetylene, $H - C \equiv C - H$, with its triple CC bond. Since the C_2 spacings in the bicarbides are similar (e.g. $YC_2 = 0.127$; $UC_2 = 0.134$) a triple bonding is expected to form in these, in which case a C_2 'molecule' in a (distorted) octahedral site would contribute 2 p orbitals (and 2 electrons) to the pd bonds with its metal neighbours, instead of the 3 p orbitals (and 2 electrons) from a single C atom in an octahedral site. An M atom would need then to provide only 2 valence electrons, instead of 4, for these pd bonds. This can be done by group III A metals and with one electron to spare for dd bonds.

This interpretation of the electronic structure of the metallic bicarbides is supported by band theory calculations (e.g. Gubanov et al., 1994) for YC_2 which show, as well as the usual low-lying 2s band, a large band of mainly p and d character centred at about 8 eV below E_F, representing both pp bonds in the C_2 units and pd bonds with the M atoms; and also a d band mainly above E_F but which starts below E_F and so represents some electrons which are in dd states.

6.3 THE UNCORRECTED COHESIVE ENERGY

The method of Chapter 5 is now extended to all group IV A and V A metals. For the uncorrected energy of the first group eqn. 5.1 is used with appropriately changed coefficients. For the second group the equation used is

$$U_{unc} = A' x e^{-p'R} - B' x e^{-q'R} + A e^{-pR} - B(1 - 0.8x) e^{-qR}. \qquad (6.1)$$

The data used to evaluate these coefficients, with $p/q = 4$ and $p'/q' = 3$, are given in Table 6.1. The difference in intermetallic spacings, R, of the carbide and pure metal is noteworthy. Only about 0.01 nm for group IV A, it is some three times larger for group V A. This stems partly from the stronger dd bonding (Friedel, 1969) in the group V A metals, due to their larger bond order, and partly from the weaker M–C bonding in their carbides, as indicated by their smaller heats of formation, ΔH (Table 2.1).

The cohesive parameters, in the equations for U_{unc}, are thus obtained as in Table 6.2. The values then deduced of R_{unc} and U_{unc} for other compositions, MC_x, are shown in Fig. 6.2 (lower curves) and given in Table 6.3. Also shown in Fig. 6.2 are values of R_{corr}, which take account of the tendency of the carbon cells to contract and the vacant ones to expand. They compare reasonably with experimental trends.

Table 6.1 Input data

	R	$-U$	ΔH	K
Ti	0.293	4.85		656
TiC	0.3062	14.14	1.91	1500
Zr	0.320	6.3		563
ZrC	0.3325	15.72	2.04	1387
Hf	0.316	6.42		688
HfC	0.328	15.97	2.17	1500
V	0.262	5.29		1062
VC	0.2946	13.73	1.06	2256
Nb	0.285	7.45		1070
NbC	0.3161	16.29	1.46	1850
Ta	0.286	8.11		1227
TaC	0.315	16.97	1.48	2156

Table 6.2 Parameters in the equations for the uncorrected cohesive energy

	A	B	A'	B'	p	q	p'	q'
TiC	17900	66.28	8810	242	31.78	7.94	24.1	8.03
ZrC	11759	72.6	14457	311	26.97	6.74	23.41	7.81
HfC	22820	87.1	13313	306	29.35	7.34	23.53	7.84
VC	25683	77.5	51546	394	36.59	9.15	30.69	10.23
NbC	24016	98.5	60144	464	32.2	8.05	28.68	9.56
TaC	35931	116.1	32558	388	33.2	8.30	26.80	8.93

Table 6.3 Cohesive energies, uncorrected, of non-stoichiometric carbides with NaCl-type structures. The values are per 6 M + 1 C atoms. Bracketed values are experimentally fitted.

	MC	$M_{1.1}C$	$M_{1.2}C$	$M_{1.5}C$	M_2C	M_3C	M_6C
x	1	0.917	0.833	0.667	0.5	0.333	0.167
TiC_x	(38.39)	38.39	38.37	38.36	38.32	38.29	38.24
ZrC_x	(47.22)	47.22	47.21	47.19	47.16	47.12	47.06
HfC_x	(48.07)	48.06	48.05	48.02	47.99	47.95	47.91
VC_x	(40.18)	40.12	40.04	39.83	39.52	38.99	37.74
NbC_x	(53.55)	53.51	53.42	53.27	53.03	52.64	51.94
TaC_x	(57.51)	57.49	57.38	57.22	56.98	56.64	56.06

Within each group the behaviour is very similar, but the group V A carbides show a much steeper fall of R with x. Mostly this is because of the larger difference in this group between the R values of the carbide and the pure metal, as noted earlier but, as can be seen in the Figure, a significant

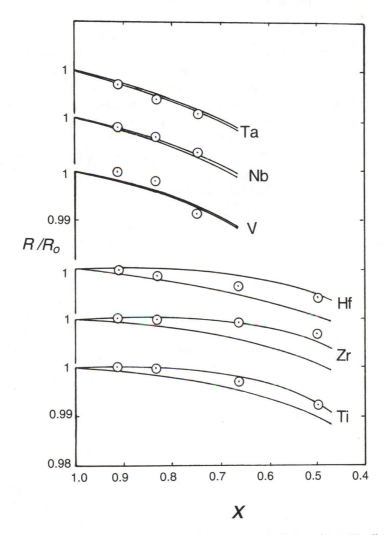

Fig. 6.2 Variation of the metal–metal atom spacing, R, in cubic MC_x alloys, relative to the value R_0 in MC. The circled points are deduced from lattice parameter values. The lower member of each pair of theoretical curves is R_{unc} and the upper one is R_{corr}.

contribution is made by the correction, $R_{corr} - R_{unc}$, which is much smaller in group VA. The reason for the latter is that, in this group, the contrast between the contractile forces in the carbon cells and the expansive ones in the vacant cells is much less pronounced. The weak contraction is due to the weaker M–C bonding and the weak expansion is because the extra dd electrons provide stronger MM bonds round the vacant cells.

This is confirmed by the calculation, as in section 5.2, of the separate forces in the metal–metal and metal–carbon networks. Whereas these do not reach zero in TiC_x at any attainable composition, they do so in, for example, VC_x at $x \simeq 0.7$. Below this carbon content the V–V links are in tension from the strong pull of the dd bonds and the V–C ones in corresponding compression, in contrast to the group IVA carbides at such compositions.

Another difference between the two groups is the quite clear drop in cohesion as x is reduced in group VA, for example by at least 0.5 eV per MC_x unit as x goes from 1 to 0.5. This is explained by the different $(1 - \alpha x)$ factors in the M–M bonding. When $\alpha = 1$ (group IVA) this factor increases by 0.5 as x goes $1 \dashrightarrow 0.5$, but when $\alpha = 0.8$ (group IVA) it increases by only 0.4 so that the correspondingly increasing M–M bonding then compensates less for the loss of M–C bonds.

6.4 CORRECTIONS

As shown by their cohesive data (Table 6.1) and U_{unc} values (Table 6.3) the group IVA metals behave so similarly in their carbide formation that titanium can be taken as representative of all. The only apparently significant difference is that α-hafnium is able to dissolve more carbon, up to $x \simeq 0.15$. This is however purely an effect of the high temperature at which α-Hf can exist, up to 2630 K. At this temperature the heat of solution, 0.43, used in section 5.5 for titanium, would give a solubility $x = 0.14$ according to eqn. 5.15.

Several corrections to U_{unc} have to be considered for the group VA carbides. No elastic correction will however be made, because the separate forces in the M–C and M–M networks are here so nearly in internal equilibrium that the correspondingly small R changes, round vacant cells, produce miniscule changes of elastic energy. There is a correction for C–C repulsion, the value of which for the stoichiometric compound is embedded in the fitted U value of this. Table 6.4 gives ΔU as the estimated C–C energy

Table 6.4 Cohesive energy of one unit, M_6C, of cubic, random, vanadium carbide, given in various forms $M_yC + (6 - y)M$, where $xy = 1$. The energy due to C–C interactions is ΔU; that from the shift in Fermi level is ΔE; those from the mixing entropies are $-TS_1$ for the carbide and $-TS_2$ for the solution of C in primary V. The total, $-G$, is also given for niobium and tantalum carbides.

x	1	0.833	0.75	0.667	0.5	0.333	0.167
R_{unc}	0.2946	0.2937	0.2922	0.2910	0.2890	0.2840	0.2770
$-U_{unc}$	40.18	40.12	40.04	39.83	39.52	38.99	37.74
$-\Delta U$	0	0.04	0.06	0.08	0.12	0.16	0.21
$-\Delta E$	0	0.36	0.50	0.62	0.77	0.97	1.11
$-TS_1$	0	0.09	0.11	0.15	0.22	0.31	0.43
$-TS_2$	0.14	0.13	0.13	0.12	0.11	0.08	0
$-G(V)$	40.32	40.74	40.84	40.80	40.74	40.51	39.49
$-G(Nb)$	53.69	54.02	54.18	54.19	54.17	54.05	53.55
$-G(Ta)$	57.65	57.98	58.14	58.14	58.12	58.05	57.67

of VC_x relative to that of VC. Also given are $-TS_1$ and $-TS_2$ from the mixing entropies at the eutectic temperature, 1900 K.

The main correction in Table 6.4, absent from the group IV A carbides, is ΔE due to a drop in the Fermi level as x is reduced. As discussed in section 4.3.2 based on Fig. 2.4, it is a consequence of the development of additional peaks in the density of states distribution, just above and below the minimum (cf. Fig. 2.1), as x is reduced. Thus, electrons which occupy the high energy levels, above the minimum in stoichiometric group V A carbides, drop down into these new peaks and thereby lower the Fermi level. The values of ΔE in Table 6.4 were calculated for $VC_{0.75}$, based on the band calculations of Redinger et al., (1985) and scaled to other compositions in proportion to the fraction of vacant carbon sites (Cottrell, 1995). The large correction which they provide is mainly responsible for the larger heat of formation of M_2C, than MC, in Table 2.1.

6.5 STABILITIES OF GROUP V A CARBIDES

The G values in Table 6.4 show that this Fermi level effect shifts the optimum composition, for carbides in the presence of excess metal, from $x \simeq 1$ (uncorrected) down to the range $x = 0.667$ to 0.75. In practice the NaCl-type phase in these carbides extends to about $x \simeq 0.6$, where it then comes into equilibrium with the M_2C phase with a CPH metal sublattice, as discussed below.

To examine the trends at the carbon-rich phase boundary, Table 6.5 gives the energies estimated for one VC unit in the forms $VC_x + (1 - x)C$, taking $U = 7.38$ per atom for carbon. This shows that the optimal cohesion occurs at about V_6C_5, which is consistent with experimental observations that vanadium does not form stoichiometric carbide. For niobium and tantalum the optimum lies nearer $x = 1$, as observed.

For the group V A carbides of M_2C type, with a CPH metal sublattice, the above contributors to cohesion carry over largely as before, but there are two significant effects. First, the contribution of the M–M interactions in the (near-ideal) CPH metal sublattice is expected to be slightly different. Because of the draining away of valence electrons to form the M–C bonds, the electron concentration for dd M–M bonding is low. The tight-binding theory of transition metals (e.g. Pettifor, 1983; Legrand, 1984) shows that at low electron concentrations the CPH structure of the metal is preferred to the FCC. Its slight energy advantage, at maximum, is about 0.05 per metal atom. Accordingly, an advantage of 0.1 per M_2C unit, for the hexagonal structure, will be assumed here.

Second, the C–C energy is changed for the reason discussed in section 5.6. According to eqn. 4.9 the extra repulsion, due to the short C–C links in the simple hexagonal C-site sublattice of the disordered M_2C structure, amounts to 0.11, 0.07, and 0.07 eV per carbon atom, respectively, for V_2C, Nb_2C and Ta_2C. These figures, added to the 0.1 advantage above, and the $-G(M_2C)$ values in Table 6.4, give $-G = 40.73$ (V_2C), 54.20 (Nb_2C) and 58.15 (Ta_2C), which are very close to the optimal values in the Table for the cubic structure. Although well within the calculational errors they show that the two phases have nearly equal energies, consistent with the observed near closure of the two-phase field, MC + M_2C, as the peritectic temperature is approached. Clearly, the hexagonal structure becomes increasingly preferred as x drops below 0.5 and the C–C repulsive energy is thereby reduced.

Table 6.5 Cohesive energy of $MC_x + (1 - x)C$. Notation as in Table 6.4.

x	1	0.917	0.833	0.75	0.667	0.5
$-G(V)$	13.73	13.90	13.97	13.93	13.79	13.58
$-G(Nb)$	16.30	16.37	16.35	16.27	16.15	15.86
$-G(Ta)$	16.95	17.03	17.02	16.93	16.81	16.52

6.6 ORDERED STRUCTURES

The C–C energy of the M_2C phase can also be lowered by the carbon atoms taking up ordered distributions among the C-sites so as to avoid the short C–C links along the hexagonal axis. Various forms of these ordered structures are observed (Toth, 1971) as shown in Fig. 6.3. The simplest is the CdI_2-type in which the basal sheets of C-sites, forming a simple hexagonal sublattice which interleaves the CPH sublattice of the M atoms, are either completely filled with carbon atoms (filled circles in Fig. 6.3a) or are completely empty. The other ordered forms can be obtained from this by transferring an ordered selection of carbon atoms from a filled sheet up into the empty one above it. When one row in three or in two is not moved, the ξ-Nb_2C or ζ-Fe_2N-type structures are respectively formed, as in Fig. 6.3b and c. A different selection of one in three produces the ϵ-Fe_2N-type (d). All of these structures provide the same benefit in eliminating all short C–C

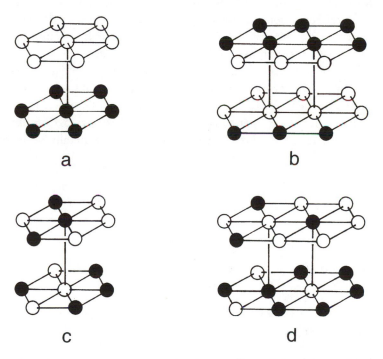

a   b

c   d

Fig. 6.3 Ordered arrangements of carbon atoms (filled circles) and vacant sites (empty circles) on the simple hexagonal C-site sublattice: (a) CdI_2-type, (b) ξ-Nb_2C, (c) ζ-Fe_2N-type, (d) ϵ-Fe_2N-type.

links. From eqn. 4.9 with $R\sqrt{2/3}$ = 0.236 (V_2C), 0.253 (Nb_2C) and 0.253 (Ta_2C) the benefit is 0.1, 0.055 and 0.055 eV per M atom respectively, for these three carbides, relative to the disordered structure.

Two other effects discriminate in energy between these ordered structures. First, the repulsion between neighbouring carbon atoms within the same basal sheet, at distance R. This, relative to the disordered structure, is greatest in the CdI_2-type structure (e.g. +0.045 eV per metal atom in V_2C, where R = 0.289); is zero for ξ-Nb_2C and ϵ-Fe_2N-type; and negative for the ζ-Fe_2N-type (e.g. −0.015 eV per M atom in V_2C). The other effect is electrostatic, as discussed in Section 5.7. The CdI_2-type structure provides that 3/6/3 distribution of M–M links of CC, CE and EE types which best enables energy to be gained by charge shift. Assuming the same value as for Ti_2C this could give the CdI_2-type structure an advantage of about 0.055 eV per M atom. An intermediate benefit is expected for the ξ-Nb_2C structure, with a 2/8/2 distribution, and none for the Fe_2N-type structures.

Consistently with this, V_2C, the small R spacing of which leads to the strongest preference for carbon–carbon avoidance, chooses the Fe_2N-type ordered structures, whereas Nb_2C and Ta_2C, the larger spacings of which reduce the C–C interactions and so should respond more to the electrostatic effect, choose the ξ-Nb_2C and CdI_2-type structures, respectively. The three effects—elimination of the short C–C bonds, number of C–C bonds in the basal sheet and the CC/CE/EE distribution—give a total ordering energy Q slightly more than 0.1 eV per M atom. On this basis, eqn. 5.17 indicates an ordering temperature of about 2000 K, in the range generally observed for these carbides.

REFERENCES

A.H. COTTRELL, *Mater. Sci. Tech.*, (1994) **10**, 788.

J. FRIEDEL, in *The Physics of Metals, I—Electrons* (J.M. Ziman ed.), Cambridge University Press (1969).

V.A. GUBANOV, A.L. IVANOVSKY and V.P. ZHUKOV, *Electronic Structure of the Refractory Carbides and Nitrides*, Cambridge University Press (1994).

B. LEGRAND, *Phil. Mag.*, (1984) B49, 171.

D.G. PETTIFOR, Chapter 3 in *Physical Metallurgy* (R.W. Cahn and P. Haasen eds), Elsevier (1983).

J. Redinger, R. Eibler, P. Hertzig, A. Neckel, R. Podloucky and E. Wimmer, *J. Phys. Chem.*, (1985) **46**, 383, 387.

L.E. Toth, *Transition Metal Carbides and Nitrides*, Academic Press, New York and London (1971).

TUNGSTEN AND MOLYBDENUM

7.1 CARBIDES OF THE GROUP VI A METALS

THE BIG structural change which occurs on moving rightwards in the Periodic Table to group VI A and beyond is the replacement of the octahedron, for the coordination of metal atoms round the central carbon atom, by the trigonal prism (cf. Fig. 1.1). Tungsten and molybdenum are transitional in providing examples of both structures in their various carbides. Associated with this change is the almost complete disappearance of the NaCl-type structure, which is so dominant in the carbides of the earlier transition metals. A phase of this type is formed at about $MC_{0.6}$ in both the tungsten and molybdenum systems, but only weakly at high temperatures. The dominating phases in these two systems, however, are $MC_{0.5}$ with a CPH metal sublattice, octahedral cells for the C-sites, an Fe_2N-type structure (cf Fig. 6.3), and some variability of composition; and the MC phase, Fig. 7.1, with a simple hexagonal metal sublattice which interleaves with an equivalent carbon one, trigonal prismatic C-cells, and with insignificant deviation from the stoichiometric composition. With regard to cohesion, an extraordinary feature of this structure is that the metallic sublattice is not close-packed (i.e. $z_{mm} = 12$), despite the fact that in these metals there are more electrons available than in the earlier ones for metal–metal bonding, but has a coordination number of only $z_{mm} = 8$.

The advantage enjoyed by this WC structure, relative to the NaCl-type, is a lower energy for the electrons in the dd M–M bonding states. This can be seen from the density of states distributions and also deduced electrostatically, as discussed below. Fig. 2.2 shows where the Fermi level would lie, at 10 valence electrons per MC unit, high in the upper d band peak, according to the rigid-band model. If Fig. 2.1 can be applied to a hypothetical NaCl-type WC or MoC, then this E_F would lie at about 13 eV above the s-band, using the latter as a fixed reference level of energy. In Fig. 2.3 however, for the WC structure, it lies at only about 11 eV above, due to the high density of d states in the region down to about 2 eV below E_F. At lower numbers of valence electrons per MC unit, as in group V A and IV A

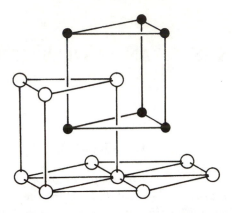

Fig. 7.1 The structure of tungsten carbide, WC, showing interleaved trigonal prismatic cells.

carbides, or in group VIA carbides with lower carbon content, this E_F advantage for the WC structure is correspondingly reduced.

The third member of group VIA, chromium, deviates more drastically from the carbide forming characteristics of the earlier groups. As noted in Chapter 1 its carbides have complex structures, low carbon contents and are all based on the cells b and c in Fig. 1.1. They will be discussed in Chapter 8. In the present chapter tungsten will be considered, but not molybdenum because its carbides are so similar to those of tungsten.

7.2 UNCORRECTED ENERGIES

For the uncorrected energies eqn. 4.7 is used, with $n = 6$, $p = 4q$ and $p' = 3q'$. Fitted to the empirical data for pure tungsten ($x = 0$), assumed to be close-packed ($z_{mm} = 12$), i.e. to $U = -8.7$, $R = 0.281$ and $K = 1875$ (eVnm^{-3})it gives

$$U_{unc}(W) = 183884e^{-39.35R} - 184.2e^{-9.84R}. \qquad (7.1)$$

The corresponding data for WC($x = 1$) are $U = -16.48$, $R = 0.287$ and $K = 2400$. Before fitting, however, the terms in eqn. 7.1 have to be modified to take account of the change to $z_{mm} = 8$ in the WC structure (Cottrell, 1995). Thus 183884 is simply reduced in proportion to the number of nearest metal atom neighbours to 122589. However, in accordance with tight-binding theory (c.f. Section 3.3), the coefficient of the bonding term is reduced only in proportion to $z^{\frac{1}{2}}$, i.e. $184.2 \to 150.4$. On this basis, the equation gives for tungsten carbide, with $z_{mm} = 8$,

Table 7.1 Cohesive parameters of tungsten carbide. The values of U are per WC_x unit.

	WC(1)	WC(2)	WC$_{0.6}$	WC$_{0.5}$
z_{mm}	8	12	12	12
R_{unc}	0.287	0.304	0.298	0.2965
$-U_{unc}$	16.48	17.19	13.61	12.67
$-\Delta E_1$	0	-1.3	-1.7	-1.4
$-\Delta E_2$	0	0	0.6	1.15
$-\Delta E_3$	0	-0.04	0.16	0.15
$-U_{corr}$	16.48	15.85	12.67	12.57

$$U_{unc} = 9695xe^{-25.9R} - 247.5xe^{-8.634R}$$

$$+ 122589\,e^{-39.35R} - 150.4\left(1 - \frac{2x}{3}\right)e^{-9.84R} \tag{7.2}$$

To obtain energies of alternative structures this can be converted to $z_{mm} = 12$ coordination which, at this uncorrected level, represents both the NaCl-type structure with its FCC metal sublattice and the WC$_{0.5}$ type with a CPH sublattice. The two M–M coefficients revert, in this, to their values in eqn. 7.1. The exponents of the two M–C terms have now to be changed because the M–C distance is $0.707\,R$ in an octahedral cell and $0.765\,R$ in a trigonal prismatic one. The equation for $z_{mm} = 12$ thus becomes

$$U_{unc} = 9695xe^{-23.94R} - 247.5xe^{-7.98R}$$

$$+ 183884\,e^{-39.35R} - 184.2\left(1 - \frac{2x}{3}\right)e^{-9.84R} \tag{7.3}$$

This has been solved for $x = 1, 0.6, 0.5$, to give the values of U_{unc} and R_{unc} in Table 7.1. They show that the uncorrected theory gives an energy advantage to $z_{mm} = 12$ (e.g. NaCl-type) WC over the hexagonal structure, $z_{mm} = 8$, of 0.71 eV per WC unit. The stability of the latter in practice thus depends on the corrections.

7.3 CORRECTED ENERGIES

It was noted in section 7.1 that the hexagonal WC structure enjoys a band structure advantage over the NaCl-type, of order of 1 eV per WC unit, due to a lowering of some of the d band energy levels. A corresponding advantage will now be deduced in terms of energies of d bonds.

Although a carbon atom has the same z_{cm} in octahedral and trigonal prismatic cells, the angular distribution of its six metal atom neighbours is different. The uncorrected theory, because it expresses energies only in terms of interatomic distances, ignores this angular effect. Assuming for simplicity that the dd bonds between metal atoms are mainly of σ character and thus directed along the lines between W–W neighbours round the C atom, the trigonal prismatic cell then has an advantage over the octahedral one in enabling the electrons of the dd bonds to avoid, more completely, the electronegative regions near the carbon atoms. The hexagonal WC structure thus satisfies, rather better, the general principle that different bonds from an atom—in this case the W–C and W–W bonds emanating from a given W atom – should point in well-separated directions.

Band theory (Liu et al., 1988) has shown that there is a net transfer of about 1.4 electrons from the tungsten to the carbon sites. On this basis an electrostatic estimate can be made, using the screened Coulomb interaction of eqn. 5.16 in the form

$$U_E = \frac{q_1 q_2}{\beta R} e^{-\beta R / r_0} \qquad (7.4)$$

where q_1 and q_2 represent respectively the charges in the C-site and those contributed by a W atom to dd bonds, βR measures the average distance between these charges and r_0 is the screening radius. It will be assumed that $q_1 = 1.4\,e$, $q_2 = 2\,e$, $\beta = 0.577$ (simple hexagonal) or 0.5 (close-packed metal sublattice), and $R = 5r_0$. Noting that a W–W bond is exposed to two neighbouring carbon interactions in the close-packed sublattice and, on average, to 2.25 in the simple hexagonal one, the deduced values of U_E then give the ΔE_1 of Table 7.1. These indicate the electrostatic disadvantage belonging to the $z_{mm} = 12$ structures. We see that in stoichiometric WC this angular effect is sufficient to make the simple hexagonal ($z_{mm} = 8$) structure more stable than the alternatives despite the less favourable uncorrected energy.

If the same comparison is made for MC carbides of the earlier transition metals, q_1 is then reduced (to about 0.5 e) and the smaller number of electrons available for dd bonding also reduces q_2, practically to zero in the case of group IV A. Thus, U_E is greatly reduced, bringing ΔE_1 down to about 0.2 for group V A and zero for group IV A, so that the unfavourable uncorrected energy of the simple hexagonal structure would then dominate, in agreement with the absence of this structure in the early transition metal carbides.

The values of ΔE_1 for the other compositions, WC_x, in Table 7.1 are deduced by using the appropriate values of R and by scaling U_E according to the factor $3x(1 - 2x/3)$ to take account of the smaller number of carbon atoms and greater number of dd electrons.

Table 7.1 also gives other corrections. For $WC_{0.5}$ these are based on the assumption that the carbon is ordered into a CdI_2-type structure (c.f. Fig. 6.3).

The presence of vacant C-sites in the non-stoichiometric carbides is expected to diminish the electrostatic disadvantage of the $z_{mm} = 12$ structures through enabling the dd electrons to shift, away from those W–W links that pass between two neighbouring C atoms ('CC-type' links), to those that pass between empty C-sites ('EE-type'). This effect, already considered for ordered TiC (cf. Section 5.8), is expected to be larger here because of the increased values of q_1 and q_2 (eqn. 7.4). However, it is not expected to raise the electron content of an EE link, which is $2\,e/3$ before redistribution, above $1\,e$, because if a second electron were to appear in an EE link it would incur the penalty of a Hubbard electrostatic repulsion, $U = e^2/r$ which, even after allowing for screening, is expected to be a few electron-volts (Friedel, 1969).

With this limitation, a maximum dd charge shift $q_2 = 0.5\,e$ per W atom will be assumed in eqn. 7.4. The shift is away from two C atoms, each with $q_1 = 1.4\,e$, at a distance of about $R/2$ from the CC-type link between them. In ordered CdI_2-type $WC_{0.5}$ there are 3 CC-type and 3 EE-type links from each W atom, in the 3/6/3 arrangement which is optimal for the maximum charge shift. On this basis, eqn. 7.4 gives an energy saving $\Delta E_2 = -1.15$ per W atom. The corresponding benefit for $WC_{0.6}$ is smaller, partly because of the disordered carbon distribution and partly because of the larger carbon content.

Finally, Table 7.1 includes an allowance ΔE_3 for C–C repulsion according to eqn. 4.9. This gives a small advantage to the non-stoichiometric carbides due to the lower density of C neighbours in these.

7.4 RELATIVE STABILITIES

The addition of these corrections produces the values U_{corr} in Table 7.1. To examine the relative stabilities of the various carbides the energies of a fixed number of tungsten and carbon atoms, in various carbide states, can now be compared for both C-rich and W-rich compositions. For the C-rich one the reactions

$$\underset{(16.48)}{WC} \rightarrow \underset{(12.57\ +\ 3.69\ =\ 16.26)}{WC_{0.5} + 0.5C} \tag{7.5}$$

$$\underset{(16.48)}{WC} \rightarrow \underset{(12.67\ +\ 2.95\ =\ 15.62)}{WC_{0.6} + 0.4C} \tag{7.6}$$

show that the stable phase is hexagonal WC, as expected. For the W-rich case the reactions

$$\underset{(25.14)}{W_2C} \rightarrow \underset{(16.48\ +\ 8.7\ =\ 25.18)}{WC + W} \tag{7.7}$$

$$\underset{(21.16)}{1.67WC_{0.6}} \rightarrow \underset{(16.48\ +\ 5.8\ =\ 22.28)}{WC + 0.67W} \tag{7.8}$$

suggest that here also WC should be the most stable, contrary to observation. However, the small difference, 0.04 eV, is well within the calculational errors. Clearly, these two carbides have nearly equal stabilities, as is shown by the observed heats of carbide formation (Table 2.1) which give W_2C an advantage over WC of only 0.12 eV per carbon atom.

The difference in energy between $WC_{0.6}$ and the alternatives is too large to be overcome by the entropic effect of a disordered carbon distribution at the temperature (3000 K) at which this phase exists. Its cohesion, like that of $WC_{0.5}$, is evidently slightly underestimated.

By replacing the corrected energy value of hexagonal WC, in eqns 7.5 and 7.7, with that of WC(2), e.g. NaCl-type, from Table 7.1 we see that if the hexagonal WC(1) did not exist there would be no stable stoichiometric carbide. The situation in the carbides of chromium and the later transition metals is thus beginning to be anticipated, here.

7.5 STOICHIOMETRY OF HEXAGONAL WC

The hexagonal WC phase contrasts strikingly with the carbides of the early transition metals (and with $WC_{0.6}$ and $WC_{0.5}$) in existing only at the sharp composition WC. This can be explained as follows (Cottrell, 1995). It is evident and also follows from the uncorrected theory that the simple hexagonal W sublattice, considered by itself, is mechanically unstable against shearing along the basal plane into a more close-packed configuration. Due to the short range of the dominating interactions in WC, the cage of W atoms round a single vacant C-site would thus tend to collapse by such a shear, although it would of course be supported against this by the elastic resistance of the carbon-containing surroundings.

Two or more such vacant sites would then attract one another since, by moving into neighbouring positions, they could mutually reinforce their individual tendencies to collapse, which could then develop more freely. This attraction would thus generate a vacancy–vacancy bond energy, which could prevail against the entropy of vacancy dispersion (except at extremely low vacancy concentrations). Vacancies would thus cluster together and so generate, locally, a close-packed phase of composition WC_x where $x \ll 1$. Thus the hexagonal phase cannot exist with x significantly less than unity since an attempt to drive down the carbon content further would, instead, cause a second phase such as $WC_{0.5}$ to form.

REFERENCES

A.H. Cottrell, *Mater. Sci. Tech.*, (1995) **11**, in press.

J. Friedel, in *The Physics of Metals, I—Electrons* (J.M. Ziman ed.), Cambridge University Press (1969).

A.Y. Liu, R.M. Wentzcovitch and M.L. Cohen, *Phys. Rev.*, (1988) **B38**, 9483.

8

CHROMIUM AND THE LATER TRANSITION METALS

8.1 THE INSTABILITY OF THE LATER CARBIDES

It was seen in Chapter 2 (Table 2.1) that the group IV A metals form the most strongly bound carbides and that the heat of formation, ΔH, weakens progressively as one moves along each row of the periodic table to the later transition metals. Broadly, ΔH drops from about 2 eV per MC unit for group IV A to about 1 to 1.5 for group V A, 0.5 for group VI A and thereafter hovers about zero or becomes negative.

On a rigid-band basis, starting from that for stoichiometric NaCl-type TiC (Fig. 2.1), this can be understood in terms of the filling of high energy states by the excess electrons of the later metals, as indicated in Fig. 2.2. The estimate of this effect (cf Chapter 6) shows that for group V A carbides such as VC it implies an energy disadvantage of about 1 eV per MC unit (Cottrell, 1995). If Fig. 2.1 could continue to be used similarly for the carbides of the later metals, then the further filling of antibonding states by the additional electrons of these would produce much more disadvantage. However, such carbides, insofar as they exist, have much lower carbon contents as well as complex structures, so that a less specific form of the argument is needed.

It was pointed out in Section 2.2 that this can be provided by considering the density of valence electrons, which increases from 8 per MC unit, for group IV A, to 9 for group V A, 10 for group VI A, and so on. Those surplus to the requirements of the C–M bonds (8 per MC unit) form dd bonds of increasing electron content in the later metals. The electron–electron repulsion between these two sets of bonds then raises energy levels, so leading to a reduced ΔH. It was also shown, in Chapter 7, that for WC in the NaCl-type structure, this could be a large effect. A simple electrostatic estimate using eqn. 7.4 and the data for WC gives this repulsive energy as about 4.3 eV per MC unit.

As indicated in Table 7.1, the conversion from the NaCl-type structure to the hexagonal one of WC, by providing a greater separation of the C–W and

W–W bonds, eliminates about 1.3 eV of this repulsive energy, which represents a significant advantage for the trigonal prismatic cell, over the octahedral one, so that there is a general preference for it in group VI A and later metals.

Clearly, the unfavourable effect of high electron concentration is accentuated in the carbides of the later transition metals. In fact, from the ΔH trends in Table 2.1, it might be expected that no carbides would be formed by any of these. The change to a trigonal prismatic cell can provide only a limited—though important—relief from the repulsion, generally insufficient to save the stability of the later carbides.

Consistent with this, no carbides are found amongst such metals in the second and third long periods. But the 3d ones do show a continuing formation, if only as metastable phases in metals such as iron, so that there lingers a slight stability here. These all have compositions and structures far removed from NaCl-type MC, however. Their low carbon content is important. If there were no redistribution of electrons among M–M bonds then the local density next to a carbon atom would of course be as adverse as in stoichiometric MC. But as we have seen in ordered $MC_{0.5}$ phases in Chapters 5 and 6, the existence of empty C-sites provides opportunity for d electrons in M–M bonds to reduce electrostatic energy by redistributing, away from carbon atoms and into 'EE' type M–M links between empty pairs of C-sites. This is one effect that favours these low-carbon carbides. Another is the reduced C–C repulsion. A third, discussed later, stems from the complexity of structure round the trigonal prismatic cells. Examples of all these effects are provided in the carbides of chromium, which occupies a unique transitional position, sharing some of the high-carbon characteristics of the early transition metals with the low-carbon contents and complex structures of the later ones.

8.2 CHROMIUM

Chromium differs strikingly in its carbide formation from its companions in group VI A. The carbide that is least dissimilar from WC is Cr_3C_2 (Fig. 1.3). Both have the metal atoms in trigonal prismatic cells and the carbon coordination $z_{cm} = 6$. The pattern of these cells is different, however, so that $z_{mm} = 10$ in Cr_3C_2 but only 8 in WC. The C–C distances are also different, all of which give advantages to the Cr_3C_2 structure.

From the observed heat of formation of Cr_3C_2, 0.295 per Cr atom (Table

2.1), and with $U = -4.1$ per atom for chromium and -7.38 per C atom for graphite, the cohesive energy of $CrC_{0.67}$ is -9.31. Thus, that of one unit of CrC in the form $CrC_{0.67} + 0.33C$ is -11.78.

To investigate why chromium forms the Cr_3C_2 structure whereas tungsten prefers WC, the first step is to express the cohesive energies of the pure metals. Using the empirical data for chromium, $R = 0.257$ (assuming $z = 12$) and $K = 1030$, eqn. 4.7 with $x = 0$ and $p = 4q$ gives for the pure metal

$$U_{unc} = 45642\,e^{-40.53R} - 73.86\,e^{-10.13\,R}. \qquad (8.1)$$

This is then applied to the Cr–Cr bonds in Cr_3C_2 by reducing the first term, $45642 \times (10/12) \rightarrow 38055$, and the second, $73.86 \times (10/12)^{\frac{1}{2}} \rightarrow 67.42$, in accordance with the drop in z_{mm} from 12 to 10; and by further reducing the second term in proportion to the reduction, from 6 to $6 - 4(2/3) = 3.33$, in the number of valence electrons available for dd bonding between the Cr atoms. The contribution of the M–M sublattice to the cohesive energy of Cr_3C_2 is thus

$$U(MM) = 38035\,e^{-40.53\,R} - 37.46\,e^{-10.13\,R} \qquad (8.2)$$

per Cr atom. At the observed average $R \simeq 0.27$ this is $U = -1.76$.

To estimate the cohesive energy of CrC in the WC structure we note first that the observed heat of Cr_3C_2 formation does not truly indicate the affinity of chromium for carbon since the U of Cr_3C_2 is diminished by the loss of M–M contribution, i.e. $-4.1 \rightarrow -1.76$. Hence the heat of the C–M sublattice in Cr_3C_2 is $0.295 + (4.1 - 1.76) = 2.64$. The corresponding value in CrC will thus be assumed to be $2.64 \times (3/2) = 3.96$, i.e. increased by 1.32 relative to Cr_3C_2. There is a compensating decrease in $U(MM)$, due partly to the decrease in z_{mm}, $10 \rightarrow 8$, and partly to the further reduction, to 2 per Cr atom, in the number of electrons in M–M bonding. These changes give $U(MM) = -0.77$, i.e. a decrease of 0.99. Thus the cohesive energy of a unit of CrC in the WC form is estimated to exceed that of $CrC_{0.67} + 0.33\,C$ by 0.33 eV.

The Cr_3C_2 structure benefits, however, from a smaller C–C repulsion. In the WC structure of CrC a carbon atom would have two (vertical) carbon neighbours at distance 0.283 and six (horizontal) ones at 0.257, giving from eqn. 4.9 $U_{rep} = 0.645$ per unit of CrC. In Cr_3C_2 it again has two vertical neighbours, at 0.283. In a vertical cell (cf Fig. 1.3) it has two more at about 0.292 and another two at about 0.300. In a horizontal cell it has two such at about 0.292 and another two at about 0.307. These interactions add up to

U_{rep} = 0.222 per $CrC_{0.67}$ unit. The C–C advantage to the latter is thus 0.42.

The reaction

$$Cr\ C(WC\text{-type}) \rightarrow Cr\ C_{0.67} + 0.33C \qquad (8.3)$$
$$\underset{(-12.11 + 0.645 = -11.46)}{} \quad \underset{(-11.78 + 0.222 = -11.56)}{}$$

is thus marginally favoured, although the estimated energy difference is small. A significant role for C–C repulsion is thus indicated, brought about by the small interatomic spacing in chromium and its carbides. By contrast, the larger spacing of tungsten and its carbides reduces the C–C role to an advantage of only 0.1 for $WC_{0.67}$ in the Cr_3C_2 structure. A repeat of the above calculation for tungsten suggests in fact that, in the equivalent to eqn. 8.3, $WC_{0.67}$ + 0.33 C would be disadvantaged by about 0.2 eV relative to WC.

Considering now the lower carbides of chromium, the requirement here for a lower density of C-cells provides opportunity for another advantage. Structures can be formed in which the number of chromium neighbours to a carbon atom exceeds six. In these carbides of complex structure and low carbon content chromium behaves like the later metals of the 3d series.

8.3   HIGHER COORDINATION IN THE LOWER CARBIDES

Despite its particularly complex structure (Fig. 1.4) $M_{23}C_6$ provides the simplest of such carbides because here all the M neighbours (8) to a C atom are at the same distance. In $Cr_{23}C_6$ this is 0.21 nm in a cell round which the M–M distance averages about $R = 0.25$. Since the ratio of these, 0.84, is larger than for a trigonal prismatic cell, 0.765, the values of the exponents, i.e. p' and q' in eqn. 4.7, need to be altered in a comparison of these two structures.

In the absence of sufficient experimental data a model system will be used, based analogously on the cohesive parameters for WC (eqn. 7.2). Thus the values $p' = 27$, $q' = 9$ will be assumed for M_3C_2 and $p' = 30$, $q' = 10$, for $M_{23}C_6$. From section 8.2 the contribution of the Cr–C network to the cohesion of Cr_3C_2 is -7.6 eV per $CrC_{0.67}$, which is consistent with the following model of the Cr–C network,

$$U = 7520x e^{-27R} - 188x e^{-9R} \qquad (8.4)$$

per $CrC_{0.67}$ unit, with $R = 0.27$ and with pre-exponential coefficients chosen to match those of WC (eqn. 7.2). This can be converted to the $Cr_{23}C_6$

network by changing p' and q'; and introducing the coordination number changes, 8/6 and $(8/6)^{\frac{1}{2}}$ respectively in the two terms. Thus for $Cr_{23}C_6$

$$U = 10030xe^{-30R} - 217xe^{-10R} \tag{8.5}$$

With $x = 0.261$ and $R = 0.25$ this gives $U = -3.20$ per chromium atom. To this must be added the energy of the Cr–Cr network, which was $U(MM)$ $= -1.76$ per Cr atom for Cr_3C_2 (eqn. 8.2). The greater number of dd bond electrons, i.e. $6 - 4(6/23) = 4.96$, in place of the 3.33 of Cr_3C_2, conveys an energy advantage to $Cr_{23}C_6$. Allowing also for its slightly larger average z_{mm} $= 11$ (Table 7.1) this carbide has

$$U(MM) = 41838\,e^{-40.53R} - 58.52\,e^{-10.13R} \tag{8.6}$$

in place of eqn. 8.2. With $R = 0.25$, $U(MM) = -2.99$ per Cr atom.

Thus the reaction

$$\underset{(-9.31)}{Cr\,C_{0.67}} \rightarrow \underset{(-3.20 - 2.99 - 3.03 = -9.22)}{Cr\,C_{0.261} + 0.41C} \tag{8.7}$$

slightly favours Cr_3C_2 in a carbon-rich environment. The corresponding estimates for a chromium-rich environment can be made from

$$\underset{(-15.8)}{Cr_{2.55}\,C_{0.67}} \rightarrow \underset{(-9.31 - 6.35 = -15.66)}{Cr\,C_{0.67} + 1.55Cr} \tag{8.8}$$

which gives a slight advantage in this case to $Cr_{23}C_6$. These results are satisfactory but of course are derived only from a model of the Cr–C network. Nevertheless, they confirm that the two carbides are closely competitive in stability.

8.4 LOWER CARBIDES WITH TRIGONAL PRISMATIC CELLS

The large square faces of the trigonal prismatic cell allow lower carbides to form complex structures in which z_{cm} is increased by the presence of nearby metal atoms outside those faces (cf Section 1.3). The M_7C_3 cell (Fig. 1.5) has one such external atom at a distance from the central carbon atom about 1.3 of the internal C–M distance. The M_3C (cementite) cell (Fig. 1.6) has two at a distance of about 1.18.

The C–M interactions are thus now complicated by the two different distances involved. This is no problem for the repulsive interaction, $A'x \exp(-p'R)$, since this is linear in z so that contributions from the various metal atom neighbours simply add, each with a p' appropriate to its distance from the carbon atom. The bonding interactions however, in $B'x \exp(-q'R)$, have a $z^{\frac{1}{2}}$ dependence. The more general equations of Pettifor (eqns 3.4–3.7) extend the $z^{\frac{1}{2}}$ analysis to situations where the individual atom–atom interactions, $h(R)$, are not necessarily equal, for example varying according to different distances R. (In those equations R is the actual interatomic distance between the two interacting atoms being considered, whereas in eqn. 4.7 and throughout this text R, the M–M spacing, is used as a general measure against which all distances are measured as fractions.)

The essential feature of those equations is that the contributions of the various bonds to an atom are given as the square root of the sum of squares of the individual bond strengths, h. In the present examples we have 6 bonding neighbours at the corners of the trigonal prismatic cell; and n more, i.e. one (M_7C_3) or two (M_3C), outside the square faces which, because of their greater distances, have weaker interactions, i.e. a fraction α of those from the cell atoms. Thus in place of the simple z_{cm} of the single-distance structures we now have

$$\bar{z}_{cm} = 6 + n\alpha^2 \qquad (8.9)$$

as an effective z_{cm} for the bonding C–M interactions.

The values of α can be estimated from the change in the value of $\exp(-q'R)$ due to increasing q' by 1.3(M_7C_3) or 1.18(M_3C). With $q' = 10$ and $R = 0.25$, as in eqn. 8.5, this gives $\alpha = 0.47$ and $\bar{z}_{cm}^{\frac{1}{2}} = 2.49$ for M_7C_3; and correspondingly 0.64 and 2.61 for M_3C. The effective contributions to cohesion of these external atoms are increased relative to those in the 8-cell of $M_{23}C_6$ because, being at relatively large distances from the central carbon atom, their repulsive interactions, $A'x \exp(-p'R)$, are correspondingly reduced. Their cohesion can thus be quite strong. For example, an earlier estimate for Fe_3C suggested that the increased number of C–Fe bonds

Table 8.1 Coordination numbers: z_{mm}, between neighbouring M atoms, averaged for complex sublattices; z_{cm}, between C and M atoms; $\bar{z}_{cm}$, ditto, defined as in eqn. 8.9.

Carbide	TiC	WC	V_2C	Cr_3C_2	Cr_7C_3	Fe_3C	$Cr_{23}C_6$
x	1	1	0.5	0.667	0.429	0.333	0.261
M subl.	FCC	simple hex	CPH	compl.	compl.	compl.	compl.
z_{mm}	12	8	12	10	10.7	11.3	11
z_{cm}, $\bar{z}_{cm}$	6	6	6	6	6.22	6.8	8

(at the expense of a slightly lower number of Fe–Fe ones; cf Table 8.1) gives the cementite structure an advantage of 0.45 eV per Fe_3C unit over alternatives (Cottrell, 1993).

8.5 NON-CARBIDE-FORMING METALS

By contrast with the 3d metals, those following molybdenum and tungsten do not form carbides. A qualitative explanation can now be suggested. As we have seen, the general decrease in carbide stability in the later transition metals is ameliorated by the formation of complex structures which allow $\bar{z}_{cm}$ > 6. However, this calls for carbon cells of the trigonal prismatic or square antiprismatic types, embedded in suitably complex crystal structures; but these cells have relatively large C-sites for a given metal–metal spacing, R (Section 1.1).

This is significant in relation to the M–M spacing, R_{mm}, at which the M–M network would be in internal equilibrium. In these cells $R_{mm} > R$ so that the M–M network is in compression and the C–M one thereby in tension, i.e. the C-site hole is too large for the C–M bonds. This can lead to an appreciable energy disadvantage. For example, eqn. 7.2 gives the equilibrium in the C–M network of hexagonal WC at R_{CM} = 0.276, whereas the observed R = 0.287. This misfit reduces the cohesion of the C–M sublattice by 0.18 eV per WC unit, according to eqn. 7.2. The reduction rapidly worsens with increasing misfit; for example to 0.75 eV for R = 0.30 and to 1.35 for R = 0.31.

It is thus important for such cells to have small misfits. This is best accomplished by small metal atoms. To illustrate the effect, we can compare R_{mm}/R for those carbides of tungsten and chromium which have such cells, as in Table 8.2, derived from the expression

$$U(\text{MM}) = \left(\frac{z_{mm}}{12}\right)Ae^{-pR} - \left(\frac{z_{mm}}{12}\right)^{\frac{1}{2}}\left(1 - \frac{2x}{3}\right)Be^{-qR} \qquad (8.10)$$

for the M–M sublattice. The M–M misfit is small for chromium, so that its trigonal prismatic and square antiprismatic cells incur little misfit penalty. Presumably the same applies to the 3d metals following chromium, since they have similarly small atoms. By contrast the misfit is much larger for tungsten, which implies a correspondingly large penalty. The elements

Table 8.2 Comparison of R_{mm}, at which the M–M sublattice is in internal equilibrium, with R, the average observed spacing, for tungsten and chromium carbides.

	WC	Cr_3C_2	Cr_7C_3	$Cr_{23}C_6$
Cell	tr. prism	tr. prism	tr. prism	sq. antip.
x	1	0.667	0.429	0.261
z_{mm}	8	10	10.7	11
R_{mm}	0.312	0.273	0.266	0.262
R	0.287	0.270	0.268	0.257
R_{mm}/R	1.09	1.01	0.99	1.02

following tungsten (and molybdenum in the 4d series) have similarly large atoms, so that their carbides would be destabilized by large misfit energies.

REFERENCES

A.H. COTTRELL, *Mater. Sci. Tech.*, (1993) **9**, 277.

A.H. COTTRELL, *Mater. Sci. Tech.*, (1995) **11**, in press.

INDEX

ALSO FROM THE INSTITUTE OF MATERIALS

Electron Theory in Alloy Design

EDITED BY D. G. PETTIFOR AND A. H. COTTRELL

'The book is well worth buying both by those involved in research and especially by final year undergraduates who I believe will be inspired by the contents.'

H. K. D. H. Bhadeshia, *Metals and Materials*

CONTRIBUTORS: A. H. Cottrell, M. W. Finnis, J. Hafner, D. G. Pettifor, F. Ducastelle, A. T. Paxton, A. P. Sutton, J. B. Pethica, H. Rafii-Tabar, J. A. Nieminen, R. Coehoorn and D. M. Edwards

Book 534   ISBN 0 901716 17 0

Introduction to the Modern Theory of Metals

A. H. COTTRELL

'Essential reading for students of metallurgy and solid-state physics' D. G. Pettifor, *Nature*

Book 403   ISBN 0 904357 97 X

Advances in Physical Metallurgy
A Collection of Papers to Mark the 70th Birthday of Professor Sir Alan Cottrell

EDITED BY J. A. CHARLES AND G. C. SMITH

CONTRIBUTORS: D. Hull, B. A. Bilby, N. J. Petch, F. M. Burdekin, P. B. Hirsch, M. J. Whelan, R. W. Cahn, R. Bullough, P. P. Edwards, B. Ralph, G. W. Greenwood, A. Kelly, A. H. Cottrell and R. W. K. Honeycombe

Book 495   ISBN 0 901462 85 3

For further information please contact:
The Marketing Department, The Institute of Materials, 1 Carlton House Terrace, London SW1Y 5DB
Telephone: +44 (0) 171 839 4071
Fax: +44 (0) 171 839 2078